Extreme Michigan Weather

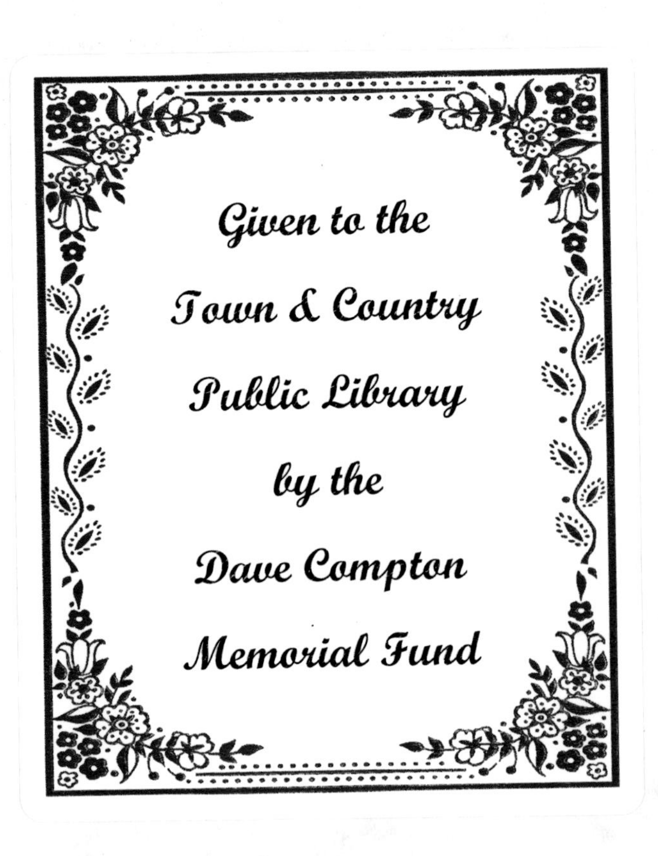
Given to the
Town & Country
Public Library
by the
Dave Compton
Memorial Fund

Extreme Michigan Weather

The Wild World of the Great Lakes State

PAUL GROSS

The University of Michigan Press
ANN ARBOR

Cover Photo: A tornado crosses Black Lake, fifteen miles southeast of Cheboygan, at 5:25 p.m. on October 18, 2007. The National Weather Service rated this twister an EF1, with winds ranging between 86 and 110 mph, and it caused scattered areas of damage along a one-eighth-mile wide, ten-mile-long path. The tornado destroyed a barn just north of M-68 and also caused tree damage. Sheri and Dr. Nathan Krinsky took this photo from the back of their home, and the author is grateful for their permission to use this spectacular photo.

Published in the United States of America by
The University of Michigan Press
Manufactured in the United States of America
♾ Printed on acid-free paper

2013 2012 2011 2010 4 3 2 1

A CIP catalog record for this book is available from the British Library.

Library of Congress Cataloging-in-Publication Data

Gross, Paul. (Paul H.)
Extreme Michigan weather : the wild world of the Great Lakes state / Paul Gross.
p. cm.
Includes index.
ISBN 978-0-472-03413-0 (pbk. : alk. paper)
1. Michigan—Climate. 2. Climatology—Michigan. 3. Weather forecasting—Michigan. I. Title.
QC984.M5G76 2010
551.69774—dc22 2010004474

Contents

Preface | vii

Acknowledgments | ix

1 Why Does Weather Happen? | 1

2 What Is "Normal Weather"? | 11

3 Severe Storms and Tornadoes | 13

4 Snow and Ice | 26

5 Water and Floods | 34

6 Heat and Cold | 39

7 Global Warming | 48

8 Miscellaneous Questions | 64

9 Michigan Weather Extremes | 72

Bibliography | 159

Index | 163

Illustrations *following page 12*

Preface

"The climate is cold and healthy. Winter sets in about the middle of November, and lasts till the middle of March, without much variation. The general face of the country is flat. Nothing like a mountain is known."

Much has changed about our knowledge of Michigan weather since this 1811 description of the Michigan Territory in *Morse's Geography.* We now know that there is tremendous variation in our weather year-round and that Michigan's changing terrain and the surrounding Great Lakes play a significant role. Anyone living a lifetime here has likely experienced intense heat waves, bitter cold, heavy snow, destructive ice storms, massive floods, high wind, severe thunderstorms, or tornadoes.

Extreme weather events are highlights in our personal time lines. Everybody remembers when an ice storm knocked out power for a week, when lightning hit a tree nearby, when a snowstorm interrupted big plans, or when record high temperatures allowed you to play some golf in January. These are things that stay in your mind forever, which is why this book is devoted to explaining the hows and whys of Michigan weather extremes. In the pages that follow, we will delve into the mysteries of extreme weather, with the hope that an increased understanding of these events will stimulate greater appreciation for Mother Nature's fury and also for the meteorologists who have to warn you about its approach.

Finally, this book includes daily weather extremes for twenty-two different cities around the state, which allow you to localize record highs, lows, and precipitation to your specific part of the state and put into proper perspective future significant weather events. I hope you enjoy reading about Michigan's weather extremes!

Acknowledgments

Many authors will tell you that writing a book can be a very challenging process. In this case, *writing* was the easy part. Having spent the past twenty-seven years as a television meteorologist broadcasting weather to large audiences and the past twenty-four years as a forensic meteorologist testifying to judges and juries about meteorological facts, words come easy to me (some family members and friends would say "too easy"). The difficult part of this book was compiling weather extremes for various cities around the state. I had to review and type 365 days of record highs, lows, precipitation, and snow for twenty-two cities into the charts you'll see later in this book. I'll save you from doing the math: this involved reviewing and transcribing well over 30,000 individual statistics—and some of them were wrong! I found occasional errors in the databases, and this resulted in much correspondence with the National Weather Service (NWS) and National Climatic Data Center (NCDC) to determine if they were indeed errors and, if so, how to officially correct them, which we did in most instances. For example, the Harbor Beach climate record had six days of record snowfall in June! NCDC checked this out for me and determined that all of these records were from June 1913 and that the statistics were all *rainfall* measurements accidentally keyed into the snowfall column in the database. They fixed the error, and the Harbor Beach climate record now correctly shows that measurable snow has never fallen in June.

I greatly appreciate the opportunity to work with Bill Deedler and Rich Pollman at the Detroit/Pontiac NWS office; Dave Berger, Gary Campbell, and James Keysor at the Gaylord NWS office; Dave Beachler, Bill Marino, and Mark Walton at the Grand Rapids NWS office; Kevin Crupi and Matt Zika at the Marquette NWS office; and Todd Holsten, Sam Lashley, and Brentley Lothamer at the Northern Indiana NWS office. I also greatly appreciate the time spent corresponding and researching with Bryant Korzeniewski, Sam McCallum, and Karsten Shein at NCDC and Sam Shey at the Midwestern Regional Climate Center; and thanks go to Linda Adams and Judith Ivan at the new and spectacular Charlevoix Public Library for their help with some historical research. Information on Great Lakes water levels came from Adam Fox and Keith Kompoltowicz at the U.S. Army Corps of Engineers, Detroit District.

Heartfelt thanks goes out to Mary Erwin, Ellen McCarthy, and Chris Milton at the University of Michigan Press. They patiently dealt with my incessant questions and guided this first-time author through the literary process to the end product you now hold in your hands. Mary, Ellen, and Chris are real gems, and I don't know how I could have accomplished this without their steering me through the uncharted waters. Special thanks also go to climate analyst Jason Samenow at the United States Environmental Protection Agency, to Dr. Jerry Meehl at the National Center for Atmospheric Research, and to Brian Montgomery at the NWS office in Albany, New York, who read my first draft and offered a great deal of helpful input. We miss you here, Brian!

I cannot adequately express my appreciation and gratitude for all the years spent working with Mal Sillars, Bob Warfield, and Chuck Gaidica at WDIV-TV. Mal brought me in as his intern back in 1981 and molded me into the meteorologist I am today; Bob is the news director who took a chance on hiring the young, raw, and not terribly good-looking kid fresh out of the University of Michigan meteorology program in 1983; and Chuck guided me through improved broadcasting skills and partnered with me on innovative projects that have earned great respect among my colleagues around the country. Most people in my business don't have the luxury of leaning on three professional mentors like Mal, Bob, and Chuck right from the start of their career.

Finally, I have to thank my family. My parents, Marion and Marvin Gross, endured raising a son that, at age seven, fantasized about becoming a meteorologist. All I talked about was the weather (with a little astronomy, archaeology, and paleontology thrown in for variety), and they fortunately didn't reroute me toward a "more sensible" career—even at a time when Dad probably hoped I would become a dentist and join him in his practice. I'll always be grateful for their support and the opportunity to pursue my childhood dream. I cannot express adequate gratitude to my wife, Nancy, and my sons, Jared and Adam. This was an extremely time-consuming project requiring many personal sacrifices, and you took the brunt of the sacrifices. I will never forget your love and encouragement throughout this difficult project.

1 Why Does Weather Happen?

Meteorology and astronomy were probably the "first" sciences, as ancient civilizations sought ways to explain weather and things they saw in the daytime and nighttime sky. Some of what they experienced was extraordinarily frightening, and there had to be an attempt to rationalize what was happening. Aristotle did just this when he wrote *Meteorologica* in 340 BCE (yes, the origin of the word *meteorology* is attributed to Aristotle!). His four-book treatise attempted to explain everything from rainbows to lightning, from wind to hail, and from comets to the salinity of the ocean. Here's how Aristotle explained hail.

> When a cloud is forced up into the upper region where the temperature is lower because reflection of the sun's rays from the earth does not reach it, the water when it gets there is frozen: and so hailstorms occur more often in summer and in warm districts because the heat forces the clouds up farther from the earth . . . But hailstones that are not rounded in shape are large in size, which is a proof that they have frozen close to the earth: for stones which fall farther are worn down the course of their fall and so become round in shape and smaller in size.

Aristotle was amazingly close in his reasoning. Storms with hail are indeed more common in the summer because the warmer, more buoyant air sometimes rises to great heights, and enormous cumulonimbus clouds—some of which rise over 50,000 feet—frequently produce hail. He also correctly deduced that the part of the atmosphere near the top of the clouds was colder and that this is where raindrops freeze and become hailstones.

There is much in *Meteorologica* that is incorrect, but you certainly cannot blame Aristotle for not understanding scientific concepts that would not be scientifically explained for two millennia. Although there were other attempts to explain the weather, Aristotle's *Meteorologica* essentially became the unquestioned meteorological authority for the next 2,000 years! Only Ptolemy's *Almagest*, a text written around 140 CE that suggested weather prediction based on the moon phase, developed any widespread influential reputation during this time.

It wasn't until the development and standardization of meteorological instruments between 1400 and 1800 CE that scientists could start quantifying various aspects of our weather. Development of the *science* of meteorology rapidly ensued, as surface observation networks recorded hourly weather, upper-air observation techniques evolved, the theory of fronts was developed, radar was applied to weather, and computers allowed scientists to model the physics of the atmosphere.

Today, we know *why* weather happens and, in particular, why many aspects of extreme weather occur. Mysteries certainly remain, and research continues, but there is one main concept to remember: **weather results from change and from rising air.** In this context, I refer to "weather" as clouds, wind, precipitation, storms, and so on, and by understanding that sharp change and enhanced velocity of rising air cause extreme weather, many of the other explanations in this book will fall simply into place.

Let's start big and then narrow our focus. As you probably already know, the earth is tilted. If you drew a line from the North Pole to the South Pole, that line is tilted 23.5 degrees from vertical. So, there are times of the year that Michigan's part of the earth is tilted toward the sun and parts of the year where we are tilted away from the sun. This leads to a trivia question: is the earth closer to the sun during Michigan's summer or winter? Believe it or not, the earth is actually slightly closer to the sun during winter! That's how important this tilt is to our weather. During summer, the Northern Hemisphere is tilted toward the sun, and the sun's rays are higher in the sky and, thus, more powerful. In winter, Michigan is tilted away from the sun, and the sun's rays are lower in the sky and weaker. If you want to see this phenomenon for yourself, stand close to a wall and shine a flashlight straight at the wall. Notice the size and brightness of the spot of light. Now, keeping the flashlight the same distance from the wall, shine it at a sharp angle. See how the light spot gets larger and dimmer? The amount of light leaving the flashlight has not changed. However, that same amount of light now has to illuminate a larger area, which dilutes the strength of the light beam. This is what happens to solar radiation in the winter.

Have you ever noticed how stormy the transitions are between winter and summer? Michigan is famous for its "Gales of November" (those strong storms that cross the Great Lakes in mid- to late fall), and we all know how spring is usually accompanied by strong storm systems and severe weather. Why does change cause stormy weather? Various laws of physics and thermodynamics state that if something changes, then something else needs to change to compensate. For example, let's look at the ideal gas law: $PV = nRT$. In this formula, P is pressure, V is volume, n is the amount of a gas, R is a gas constant, and T is temperature. So this formula says that PRESSURE multiplied by VOLUME equals AMOUNT OF GAS multiplied by a GAS CONSTANT multiplied by TEMPERATURE. This is an absolute law of physics, and it cannot be broken: the value of the two things multiplied on the left of the equal sign *must* equal the value of the three things multiplied on the right of the equal sign. So, if you have a certain amount of gas at a certain temperature and pressure, you cannot change its volume, temperature, or pressure without one of the other parameters changing to compensate. The sharper the change in temperature and pressure in the atmosphere, the stronger the resulting response is in our weather. In reality, it's a more complicated process than this because there are other laws of physics and meteorology to consider, but you get the general idea.

This leads us to the next part of our concept: rising air. Without rising air, you cannot have clouds, and without clouds, you cannot have precipitation. Here's how it works: The air you and I breathe has a certain amount of water vapor in it. When a parcel of air rises, it starts to cool as it gets higher and higher in the atmosphere (Aristotle was right—it IS cooler up there!). That parcel of air was able to hold a certain amount of water vapor at the temperature it started out at, but as it rises and cools, its capacity to hold water diminishes. Eventually, it cools to a temperature at which it cannot possibly contain any more water vapor than it already holds. This temperature is called the dewpoint temperature. When the temperature falls all the way to the dewpoint temperature, the parcel of air is 100 percent saturated with water vapor. If the temperature falls any further, then the air parcel must shed the excess water vapor. Water vapor molecules start coalescing onto little bits of dust, pollen, and pollution and form small droplets that we then see as clouds. As more and more water vapor condenses, raindrops or snowflakes develop and fall out of the cloud.

How much rain or snow falls from the cloud depends on how much water vapor is available and how fast the air is rising. Put simply, the faster and more violently humid air rises, the greater the vertical development of the clouds and the greater the chance for higher intensity precipitation. So, what factors influence how fast the air rises? Two things: the "weight" of the air and changes in the atmosphere (such as an approaching front or storm system) that dynamically *force* the air aloft. Warm, humid air is less dense than cold, dry air and, thus, is more easily lifted by approaching dynamics. Sometimes, the air is so unstable (i.e., buoyant) that weak dynamics are all it takes to get storms started. In other cases, the air is marginally unstable, but the approaching dynamics are strong enough to overcome this and violently force the air upward.

Here's an analogy that will help you visualize this: Let's say I needed a 100-pound box carried up a flight of stairs. A fellow my size (5′7″) would have trouble getting that box up the stairs. However, a Detroit Lions offensive lineman could probably get that box up there pretty easily, which is an example of strong dynamics overcoming a relatively "heavier" atmosphere. On the other hand, let's say the box I needed to take upstairs weighed only three pounds. I could run that box up the stairs myself without any trouble. This would be an example of weaker dynamics initiating lift in a very buoyant, unstable atmosphere.

Now that we have a better handle on what gets weather started, it's time to shed some light on the "triggers" that fire those atmospheric bullets upward through the troposphere. These triggers (and, yes, this term is used quite frequently in meteorological discussions) are some of the basic things we television meteorologists show you on our weather maps every day. While cold and warm fronts and high and low pressure may seem confusing to some, they are actually fairly easy concepts (see fig. 1).

To understand fronts, you need to visualize giant masses of air flowing around the planet. Imagine a cold, dry pocket of air 1,500 miles in diameter traveling across the Dakotas toward Michigan. That air mass has edges, and we're interested in the leading edge, of course, as it heads our way. Out ahead of that leading edge of cold, dry air is a southerly wind blowing warm, humid air into the state. However, this wind shifts to the northwest once the cold air arrives. The change occurs right at the leading edge of that cold air mass, and we call that leading edge the cold front. It's that easy! A cold front is simply the leading edge of colder air—the dividing line between the two air masses. Let's look in greater detail at what's happening right at the front. The northwest winds behind the front are colliding with the southerly winds ahead of the front. These winds blowing at each other from different directions converge just ahead of the front. In other words, the wind collides at that point, and since it cannot go down into the ground, it must go up. Aha! The air is now rising just ahead of the front, and if the air is unstable, it rises even more violently. In most cases, the sharper the change (there's that word again) from warm to cold, the greater the convergence at the frontal boundary. Using what we've just learned, dynamics generated by the front, combined with the properties of the air ahead of the front (temperature, humidity, etc.), dictate just how violently the air rises and, thus, how intense the resulting weather becomes. There are other factors that impact whether or not thunderstorms become severe and produce tornadoes, but this will be discussed in chapter 3.

Now that you understand cold fronts, there's really no need to spend too much time talking about warm fronts, which are simply the front edge of advancing warmer air masses. The principles are essentially the same, with the "weather" generally occurring out ahead of the front.

High and low pressure areas—H and L, respectively, on our weather maps—are also easy concepts often greatly misunderstood by most people. Air pressure is simply the weight of the column of air above you, and a barometer is like a supersensitive scale for measuring the changing weight. In the same way that the needle on your bathroom scale goes farther to the right as you gain weight (if you have a digital scale, then you'll just have to use your imagination), the needle on a barometer goes farther to the right as the atmosphere gets "heavier." Conversely, the "lighter" the air gets, the lower the barometric pressure. So, when you hear a meteorologist say that the pressure is falling, then the air is becoming more buoyant. That's why old-fashioned barometers have words like *stormy* and *rain* at the lower pressures on the dial and *fair* or *dry* at the higher pressures. If you have a barometer, you'll also notice that the faster the pressure changes, the windier it gets, and the sharper the eventual change in the weather will be.

An area of low pressure is simply the point of lowest air pressure on that part of the weather map. Air rotates counterclockwise around low pressure (in the Northern Hemisphere) and, due to surface friction, spirals in a bit toward the center of the low. As the air spins around and inward, it gets to a point where all of this buoyant air converges, and since it cannot go down into the ground, well, you now know the rest of the story—it's forced upward. That's why inclement weather generally accompanies low-pressure systems as they cross the country. High-pressure systems are simply the opposite: higher pressure is "heavier" air, with the sinking air flowing outward in a clockwise direction (in the Northern Hemisphere) from the center of the high. Since rising air is necessary for clouds and precipitation, the stable, sinking air associated with high-pressure systems generally results in fair weather.

There is one more weather feature we need to discuss: the jet stream. You have probably heard a television meteorologist talk about the jet stream on many occasions, and no, it's not just some technical thing we put on our maps to show you that we really know what we're talking about. The jet stream is a vitally important element of meteorology that we use on a daily basis to forecast both short-term and long-term weather.

What is the jet stream? It is simply the band of strongest wind aloft that slithers around the planet like a long, continuous snake with evolving peaks and valleys. If you have ever been on a passenger jet, the jet stream is roughly at the same altitude you are when the pilot announces that the plane has reached "cruising altitude."

How fast are the jet stream winds? It depends. A weak jet stream may have wind well under 100 miles per hour, while a stronger jet stream may contain wind approaching 200 miles per hour.

Why is the jet stream so important? It's the dividing line between a colder air mass to its north and a warmer air mass to its south. If you were paying attention earlier when I wrote about weather resulting from change in the atmosphere, then you suddenly just had a brilliant revelation: if the jet stream represents a large-scale area of atmospheric change, then it is also an area of generally unsettled weather—which it is! In fact, the jet stream serves as the track on which our never-ending train of storm systems travels. The jet stream's strength, combined with the depth of its peaks and valleys, dictate the birth, death, strength, and movement of all of those triggers we discussed earlier.

There are many things that force changes in the jet stream's configuration, such as the Arctic Oscillation, the North Atlantic Oscillation, the Madden/Julian Oscillation, etc., but the ones most familiar to you are probably El Niño and La Niña . . . eastward or westward shifts of warmer than normal tropical Pacific Ocean waters. There is a significant interaction between the oceans and our atmosphere, and strong El Niños and La Niñas dramatically impact Michigan winters. In general, strong El Niños give us milder, less snowy winters (such as the winter of 2009–10), while La Niñas give us colder, snowier winters (such as the winter of 2007–8).

So now you understand the three-dimensional nature of the atmosphere that challenges meteorologists every day. In order to forecast weather that will be happening where YOU care about it (at the earth's surface), meteorologists have to analyze temperature, pressure, wind, and humidity from the surface to tens of thousands of feet above the ground to determine just how extreme the resulting weather will be. For example, you have probably experienced a long stretch of hot, humid weather, with seemingly endless sunny days, except for one day in which strong thunderstorms erupted despite the fact that no cold front crossed the area. What happened that made that day different from all of the other days? It may have simply been a slightly cooler pool of air moving overhead at about 20,000 feet that increased atmospheric instability and allowed air to rise high enough to form the storms.

How do meteorologists know the weather aloft? Twice a day, at the exact same times every day at weather offices around the world, balloons with weather instruments are launched. The instrument package contains a radio transmitter that radios the data back to a ground station as the balloons float higher and higher. There are certain specific pressure levels, called "mandatory levels," that we get data from every time, in addition to data from various other levels in between. Twice a day, we get to see upper-air maps from those mandatory levels, which give us a great deal of information about the meteorology aloft. As you will see later, features such as the location and thickness of an above-freezing layer of air in an otherwise below-freezing atmosphere, the changing strength and direction of the wind as you go farther aloft, and the location of dry or humid air aloft are vitally important to meteorologists. Put simply, this upper-air data provides a vertical profile of the atmosphere that allows us to forecast precipitation type and intensity.

Finally, all of this upper-air data is ingested into some of the most complicated computer programs you'll ever hear about, to produce computer prognostications of atmospheric changes. A discussion about how meteorologists forecast the weather would fill up another book, but here's how I do it: when I walk into my weather office, I start by hand analyzing a regional surface map to get a grip on any approaching small-scale triggers or features. I then look at satellite and radar images—I call this "looking out the window"—to see what weather is happening right now. Next, I study upper-air maps to see what might be causing this weather, and then I check the computer models. I approach every weather forecast as a problem, and the various computer models offer solutions to that problem. It is important to understand that there are many different computer models to look at, and some of them are more right or wrong on some days than others. So, I have to determine which computer models are best modeling what is happening right now and are showing the most consistent handling of our approaching weather system over the past day or two of computer runs. Then I have to see if there is any overwhelming consensus in the computers' handling of the evolving weather pattern. This process results in my choosing a model or models, which forms the basis for my forecast.

The final component to making a weather forecast is experience. No matter how well they taught me theoretical meteorology at the University of Michigan, college cannot teach experience. Over my twenty-seven-year career, I have seen how computer models handle different types of storm systems and extreme weather events, and there is no substitute for this experience. Occasionally, I "throw out" the computer models and make a forecast based on past experience only. Mal Sillars used to call this "forecasting by the seat of your pants," and on these days, it is really a matter of going with a gut feeling, rather than what a computer says. As great as our computer models are, a computer lacks one essential function that humans have: the ability to reason.

Today, meteorologists are correct most of the time, which is a far cry from weather forecasts even a generation ago. For example, a five-day forecast today is as accurate as a three-day forecast was back in the 1970s. That's incredible improvement. So why are weather forecasts still occasionally wrong? Most often, either the wrong computer model was chosen as the basis for the forecast, or the computer models themselves were wrong from bad data or for other reasons.

One of the worst forecasts I ever made was for a six- to ten-inch snow when only three to six inches eventually fell. After the event, I looked back at the computer models, and there was not a single thing I would have done differently. I couldn't figure out what went wrong. So I called my friend Brian Montgomery, who was working the same shift at my local National Weather Service office. He had made a similar forecast and was looking into the same problem. He determined that poor upper-air balloon coverage just north of the Great Lakes resulted in our computer models grossly underestimating the depth of a dry layer several thousand feet aloft. Later research also showed much greater atmospheric instability up north than was projected by the computers, which further compounded the problem. As a result, the snow arrived right on schedule but evaporated before reaching the ground. It wasn't until the dry layer fully saturated that the snow started reaching the ground, and by that time, half of the moisture available for generating the snow had already been "spent."

I'll never forget that snow event, and I'll especially remember it the next time a snowstorm approaches at a time when we have an air mass flowing down from that same source region.

2 What Is "Normal Weather"?

I hate the word *normal.* I really do. It is a standard statistic used by meteorologists worldwide but, potentially, is one of the most misunderstood words ever. Most people think that "normal" is what the weather is *supposed* to be. This could not be further from the truth. "Normal," as used by meteorologists, is a mathematical statistic—no more, no less. Here's how it works, and to keep this simple, let's just consider high temperatures: At the beginning of each decade, meteorologists take a city's January 1 high temperatures for each of the past thirty years, add them up, and then divide that number by thirty. This gives us the *average* high temperature for January 1. The process is then repeated for January 2, 3, 4, and so on through December 31. Those daily average high temperatures are then plotted on a graph. If you just connected the dots, you would end up with a jagged line. Instead, a smooth line is then fitted through the dots. Each day's *normal* high temperature is taken from the smooth line on the graph.

Take December 25 as an example. Detroit's normal high is 33 degrees, but this is not what the high temperature is *supposed* to be that day. As you now know, 33 degrees is roughly the average of the past thirty years' high temperatures, and this includes a high temperature of 64 degrees in 1982, a high of 4 degrees in 1983, and everything in between. So join me in cringing the next time you hear a television weathercaster say, "Today's high was 5 degrees colder than it's supposed to be." No, "normal" is NOT what it is supposed to be. There is tremendous variety in weather, and this variety is "smoothed" by the thirty-year average so that meteorologists have a gauge to monitor our changing climate. Thirty years is long enough to give us a nice average but short enough to show changes, such as global warming.

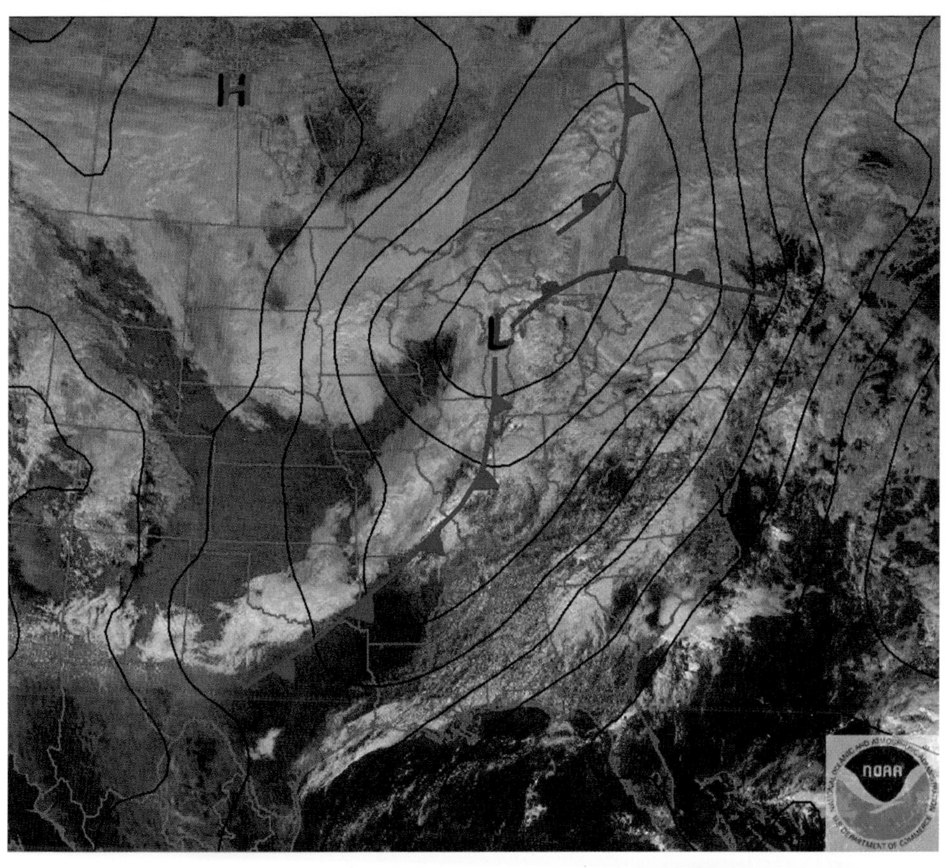

Fig. 1. A NOAA satellite image with surface features and isobars (pressure lines)

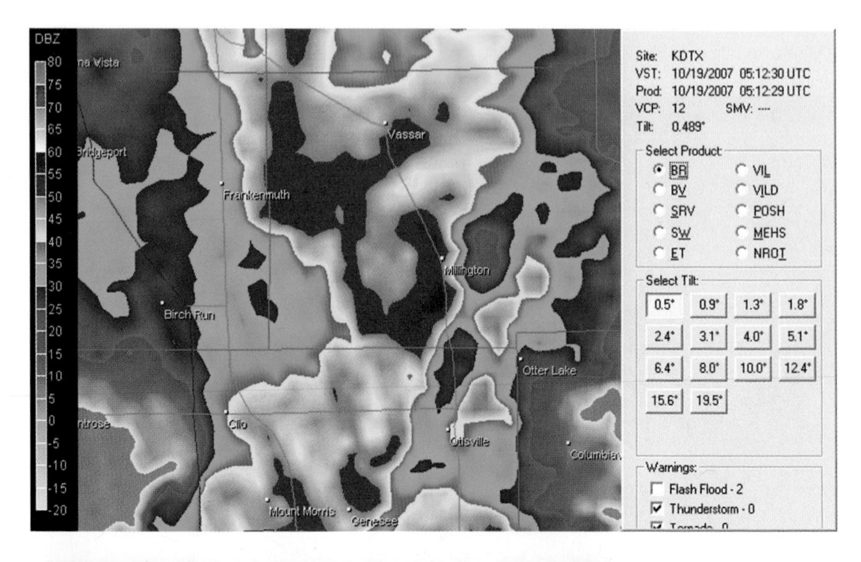

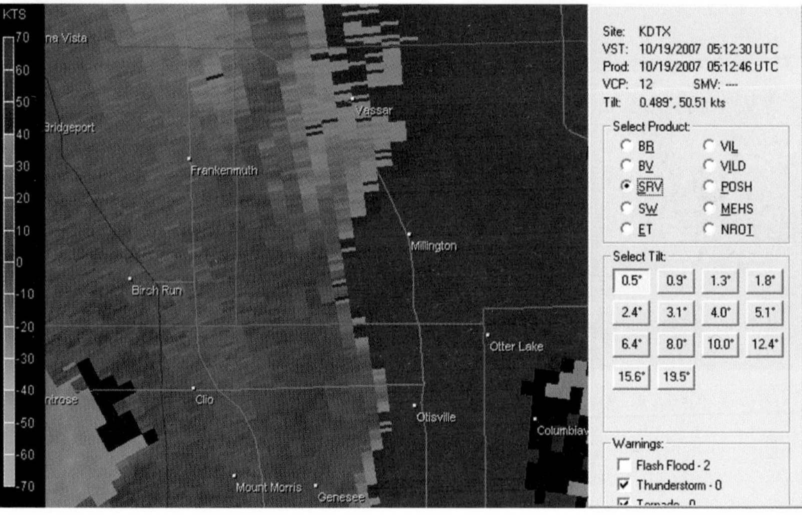

Fig. 2. National Weather Service radar images showing a tornado approaching Millington in Tuscola County at 1:12 a.m. on October 19, 2007. The top image is of base reflectivity, which shows precipitation intensity only. It would be difficult to detect the tornado based on this image alone. The bottom image is of storm relative velocity, which shows relative wind direction and speed. The radar site is located south of the lower right part of the image. Red colors indicate wind blowing away from the radar, while greenish colors indicate wind blowing toward the radar. Notice the patch of green located right next to the bright patch of red. This is an area of strong wind blowing toward the radar next to strong wind blowing away from the radar: you can actually infer the tornado's counterclockwise rotation. This data allowed the NWS to detect the tornado before it struck Millington, at a time of day when visual identification is extremely difficult.

Twister Slams into Massachusetts, Killing 69

The Detroit Free Press

209 DIE IN TORNADOES; FLINT TOLL MAY SOAR

Worcester Area Lists 738 Injured

Stricken City Digs 112 Bodies From Ruins

Many Still Missing; Damage Is 15 Million

Fig. 3. A *Detroit Free Press* front page emphasizes the disaster that unfolded on June 8, 1953, when a massive tornado devastated Beecher. This tornado is still the last single tornado in America to kill more than 100 people and ranks as the ninth deadliest tornado in U.S. history.

Fig. 4. Aerial photo of a tornado path through a Kalkaska County forest on October 18, 2007. (Photo from National Weather Service and Civil Air Patrol.)

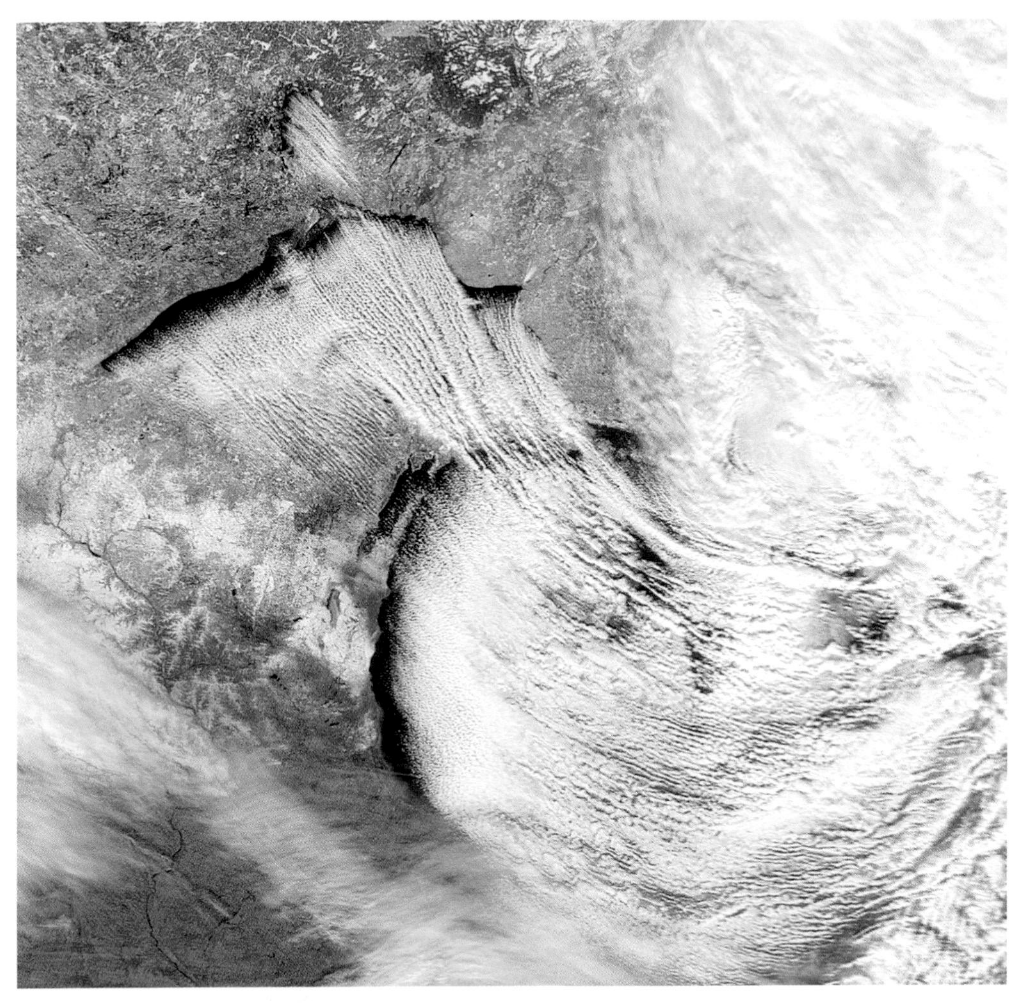

Fig. 5. Strong, cold northwest winds screaming across the relatively warmer Great Lakes generate intense lake-effect snow squalls on December 5, 2000. (Source: NASA/SeaWiFS.)

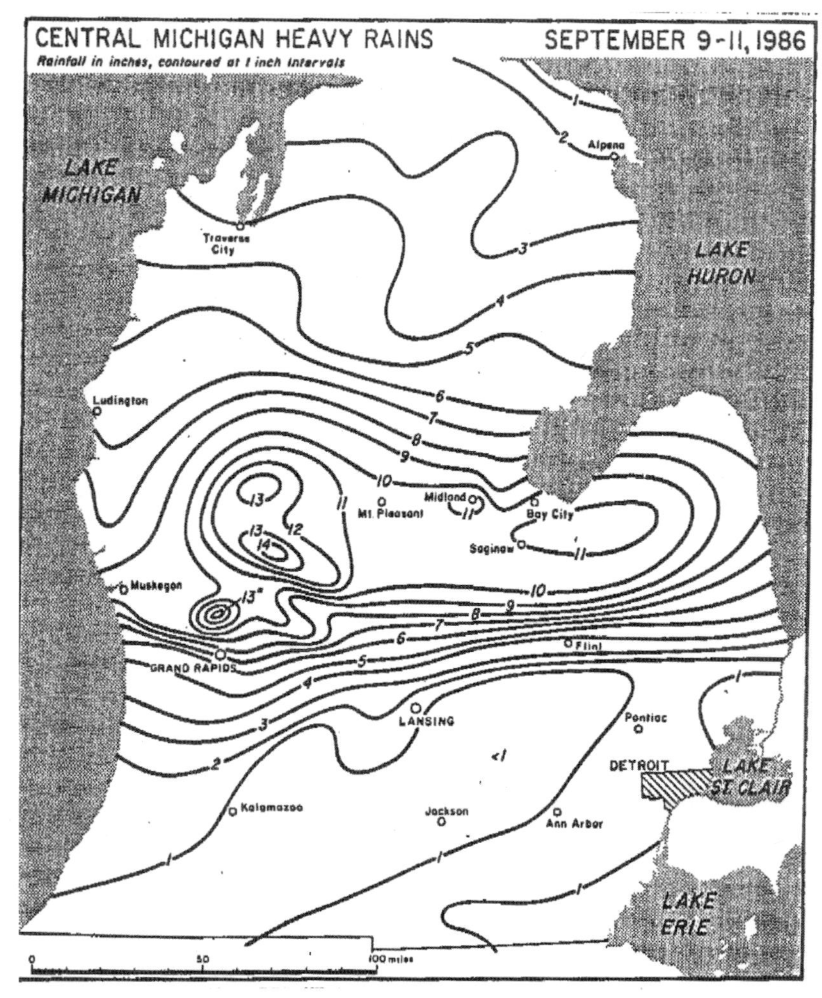

Fig. 6. Incredible rain totals from September 9–11, 1986. (Source: National Weather Service.)

Fig. 7. Freezing rain occurs when rain falls into subfreezing temperatures at the surface and freezes on contact. (Photo by Paul Gross.)

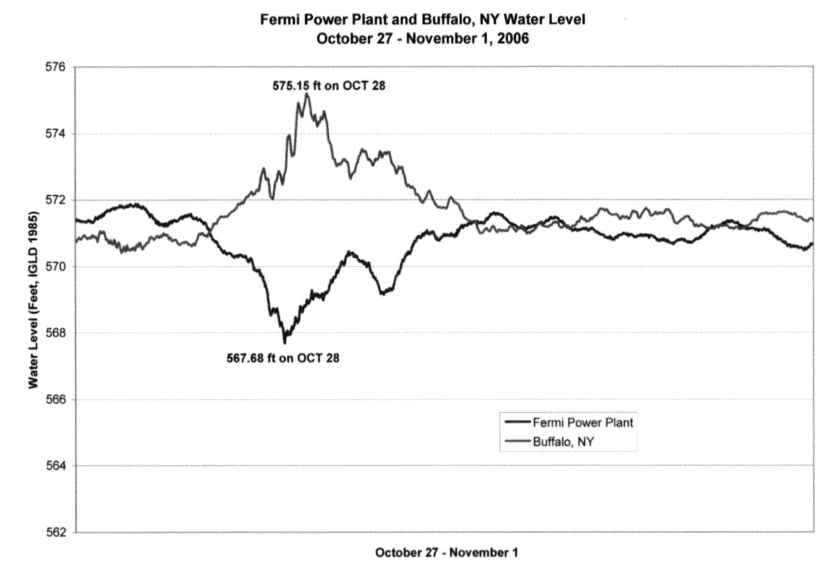

Fig. 8. Graph showing simultaneous water levels at Fermi II Nuclear Power Plant on the west shore of Lake Erie and at Buffalo, New York, during the seiche of October 28–29, 2006. (Data from NOAA; graph by United States Army Corps of Engineers, Detroit District.)

Detroit Free Press

On Guard for Over a Century — 24 Pages — Three Cents

FINAL EDITION

Calls Talk Issue

Cooling Rains Slowly Moving Eastward but No Relief Is Expected Here Today; Detroit Heat Toll Mounts to 186 Dead

Heat Deaths Pass 1,700 Total in U.S.

Government Expects to Start Purchasing Cattle This Week

Grain Prices Holding Steady in Chicago Pit

(By the Associated Press)

Rainfall in the northwest sector of the drought land started cooler weather on a new march across the sun-baked plain states Monday night as renewed onslaughts of heat to the East continued to break records and piled up the death toll.

Good rains struck sectors of the drought-parched Dakotas and Western Minnesota Monday night, lowering temperatures and signalizing

Wallace Assails Alarmists for Scare on Food Scarcity

COLORADO SPRINGS, July 13—(A. P.)—Henry A. Wallace, secretary of agriculture, lashed out tonight at "alarmists and propagandists" who, he charged, "have tried for their own purposes to scare the consumer about food scarcity" as a result of the drought.

Reciting steps the Government contemplates to aid drought-harassed farmers, particularly in the Middle West, Wallace, in a radio address said that purchases of livestock for which no feed was available will be handled so that neither consumers nor growers are penalized.

"There is no likelihood now of anything approaching a national food shortage," Wallace said.

"We are much better supplied with food for livestock than we were in 1934. There is an ample supply of vegetables and other food.

"Even if the corn crop in the Middle West should be cut severely, the effect would not be felt immediately. Any shortage in the grain inevitably reduces the pork supply a year later but on the whole there is no excuse for substantial increases in food prices now. The persons who are using the drought as an excuse to increase their profits are taking advantage of human suffering."

Prepared for Relief

Summarizing the drought situation, Wallace, who is vacationing in the Pike's Peak region, said that the Government "was prepared to take care of the drought in 1934—and it is prepared this year."

The Secretary pointed out that Congress appropriated about

State Fatalities Pass 400 on Sixth 100-Degree Day

Morgue Scene One of Catastrophe; More Than 80 Eloise Patients Die; Factories Consider Closing

With the total heat deaths in Michigan mounting past the 400 mark, this state led the other 47 Monday in the number of sun victims, and Detroit suffered its sixth consecutive day of 100-degree temperatures. The top figure was 102, at 2:30 p. m.

Deaths in Detroit mounted rapidly during the day, and at midnight the figures stood at 64, bringing the total for the present heat wave to 186. In all of Michigan, including

Fig. 9. *Detroit Free Press* front-page coverage during the Heat Wave of 1936. Detroit endured seven consecutive days of triple-digit heat.

Fig. 10. Photo of Earth from space. (Source: NASA.)

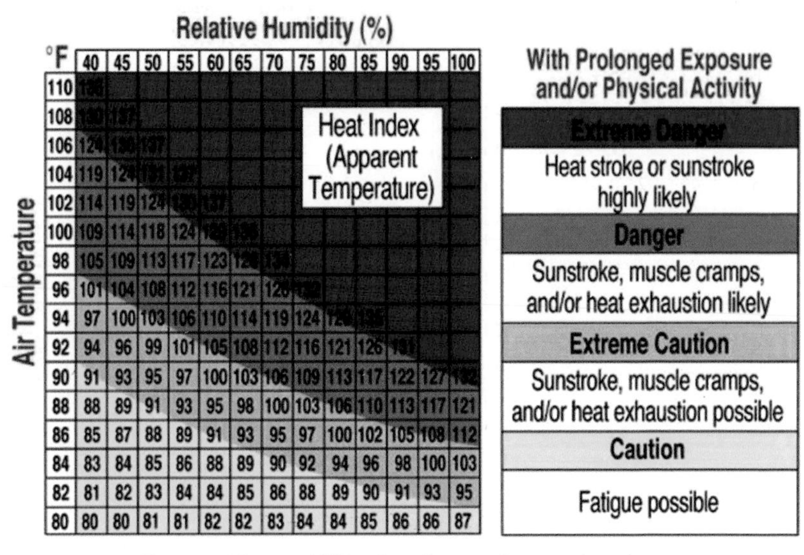

Relative Humidity (%)

°F (Air Temperature)	40	45	50	55	60	65	70	75	80	85	90	95	100
110	[illegible]												
108	[illegible]	137											
106	124	[illegible]	137										
104	119	124	[illegible]	137									
102	114	119	124	[illegible]	137								
100	109	114	118	124	[illegible]	[illegible]							
98	105	109	113	117	123	[illegible]	[illegible]						
96	101	104	108	112	116	121	[illegible]	[illegible]					
94	97	100	103	106	110	114	119	124	[illegible]	[illegible]			
92	94	96	99	101	105	108	112	116	121	126	[illegible]		
90	91	93	95	97	100	103	106	109	113	117	122	127	[illegible]
88	88	89	91	93	95	98	100	103	106	110	113	117	121
86	85	87	88	89	91	93	95	97	100	102	105	108	112
84	83	84	85	86	88	89	90	92	94	96	98	100	103
82	81	82	83	84	84	85	86	88	89	90	91	93	95
80	80	80	81	81	82	82	83	84	84	85	86	86	87

Heat Index (Apparent Temperature)

With Prolonged Exposure and/or Physical Activity
Extreme Danger
Heat stroke or sunstroke highly likely
Danger
Sunstroke, muscle cramps, and/or heat exhaustion likely
Extreme Caution
Sunstroke, muscle cramps, and/or heat exhaustion possible
Caution
Fatigue possible

Fig. 11. National Weather Service heat index chart

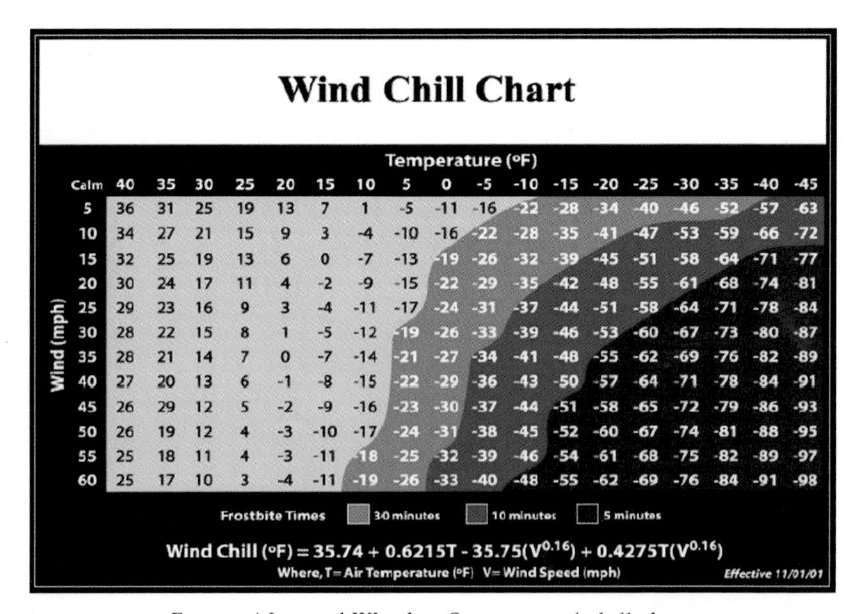

Wind Chill Chart

Temperature (°F)

Wind (mph) \ Calm	40	35	30	25	20	15	10	5	0	-5	-10	-15	-20	-25	-30	-35	-40	-45
5	36	31	25	19	13	7	1	-5	-11	-16	-22	-28	-34	-40	-46	-52	-57	-63
10	34	27	21	15	9	3	-4	-10	-16	-22	-28	-35	-41	-47	-53	-59	-66	-72
15	32	25	19	13	6	0	-7	-13	-19	-26	-32	-39	-45	-51	-58	-64	-71	-77
20	30	24	17	11	4	-2	-9	-15	-22	-29	-35	-42	-48	-55	-61	-68	-74	-81
25	29	23	16	9	3	-4	-11	-17	-24	-31	-37	-44	-51	-58	-64	-71	-78	-84
30	28	22	15	8	1	-5	-12	-19	-26	-33	-39	-46	-53	-60	-67	-73	-80	-87
35	28	21	14	7	0	-7	-14	-21	-27	-34	-41	-48	-55	-62	-69	-76	-82	-89
40	27	20	13	6	-1	-8	-15	-22	-29	-36	-43	-50	-57	-64	-71	-78	-84	-91
45	26	29	12	5	-2	-9	-16	-23	-30	-37	-44	-51	-58	-65	-72	-79	-86	-93
50	26	19	12	4	-3	-10	-17	-24	-31	-38	-45	-52	-60	-67	-74	-81	-88	-95
55	25	18	11	4	-3	-11	-18	-25	-32	-39	-46	-54	-61	-68	-75	-82	-89	-97
60	25	17	10	3	-4	-11	-19	-26	-33	-40	-48	-55	-62	-69	-76	-84	-91	-98

Frostbite Times: 30 minutes, 10 minutes, 5 minutes

Wind Chill (°F) = $35.74 + 0.6215T - 35.75(V^{0.16}) + 0.4275T(V^{0.16})$

Where, T= Air Temperature (°F) V= Wind Speed (mph)

Effective 11/01/01

Fig. 12. National Weather Service wind chill chart

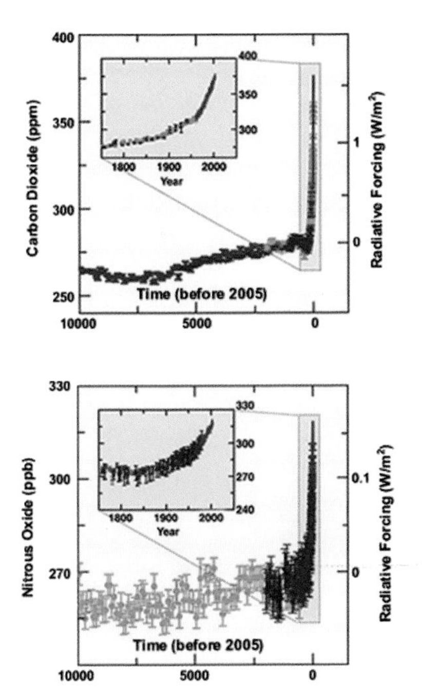

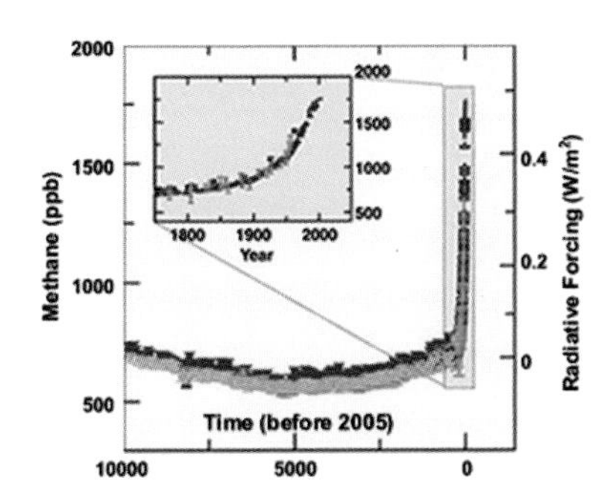

Fig. 13. Graphs showing levels of carbon dioxide (*upper left*), methane (*upper right*), and nitrous oxide (*bottom*) in Earth's atmosphere since the last ice age. Notice the relatively stable levels over the past 10,000 years, followed by an exponential increase as we approach the present. There is no disputing that human activity has significantly changed the composition of Earth's atmosphere. (Source: IPCC/COMET.)

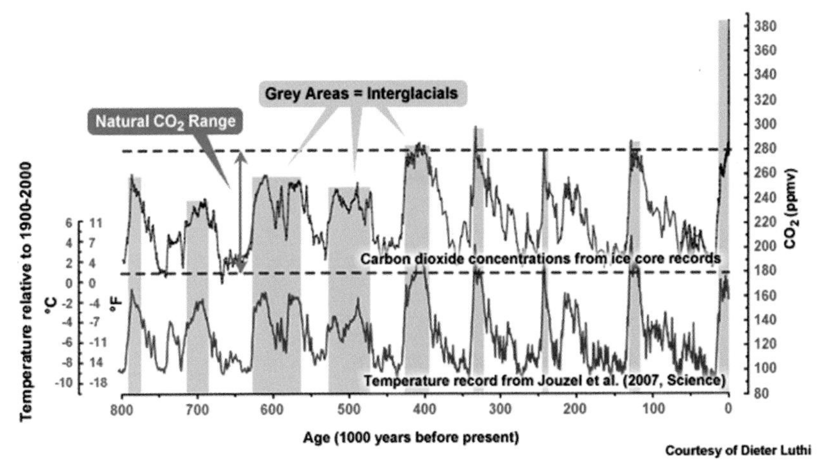

Fig. 14. Graphs of atmospheric carbon dioxide (*blue line*) and global temperatures (*red line*) for the past 800,000 years. Notice how every increase in carbon dioxide is accompanied by a corresponding increase in global temperatures. All of the significant warm periods prior to our present warming were initiated by changes in the earth's orbit or tilt, which caused a release of greenhouse gases, which further accelerated the global warming. Scientists are certain that Earth's current warming is not the result of changes in our planet's orbit or tilt. (Source: Dieter Lüthi/COMET.)

Fig. 15. Comet McNaught graces the evening skies over Hudson, Michigan, in January 2007. (Photo by John Kirchhoff.)

Fig. 16. Rainbow over Troy, Michigan. (Photo by Sandra Gross.)

Fig. 17. The author in front of the green screen (chroma-key) in the WDIV-TV studio. Notice the television monitor that also shows the weather map in the background.

Fig. 18. Warm air flowing over a cold snowpack is one way that dense fog develops. (Photo by Paul Gross.)

Fig. 19. Sightings of the aurora borealis (northern lights) are most common in northern Michigan but occasionally make their way downstate. Here, an aurora is seen over Hudson, Michigan. (Photo by John Kirchhoff.)

3 Severe Storms and Tornadoes

The following story about a Michigan tornado is true and was personally verified by this author.

> September 14, 1990: Cyndi was headed home after running a few errands with her young daughter. Her final stop was to drop by a friend's house. After pulling into her friend's driveway, Cyndi walked around her van and saw what looked like a tornado in the field behind the house across the street. She ran into the open garage and pounded on the door. "Linda, is that a tornado?" "I don't know, I've never seen a tornado!" replied Linda. "Well, I'm not sure, but I think that's a tornado!" screamed Cyndi. "We'd better get to the basement!" So Cyndi ran back to the car and grabbed her daughter, while Linda swept up her sleeping son from his crib, and they ran down into the basement. Once downstairs, they saw through one of the small windows at the top of the basement walls that the tornado was approaching the house. The two women and their children huddled in a corner of the basement, and suddenly the small windows blew out and the basement filled with dust. After what seemed like an eternity of silence, they decided to go upstairs and

see what happened. They turned the corner to walk up the basement stairs and saw that they were looking at the sky through a severely damaged roof, with boards and pieces of insulation strewn about. The house took a direct hit: the tornado's wind was strong enough to push the entire chimney into the backyard. The most remarkable aspect to this story is time: this tornado hit only thirty seconds after Cyndi and Linda reached the basement. Any delay in taking cover could have been disastrous.

Tornadoes are, without question, nature's most violent phenomenon. It takes very special meteorological ingredients to create the unique thunderstorm that generates tornadoes, and not all of these storms even drop a twister.

Most tornado days start with hot, humid weather, which makes the air very buoyant as we discussed earlier. While this makes the air unstable, having cooler, drier air aloft on top of the warm, moist air makes it even more unstable. Warm, moist air rising into the cooler, drier air starts accelerating. The cooler it is aloft, the faster the air rises, and the greater the vertical development of the cumulonimbus cloud will be. Intense downpours develop, and so does lightning.

Lightning

Lightning results from a buildup of negative charges at the base of a cloud. As the cloud moves, positive charges here on earth pool beneath the cloud. Since opposite charges attract, an invisible "leader" reaches down from the base of the cloud toward the pool of positive charges. Meanwhile, a leader of positive charges reaches upward from the ground. These positive charges flow through trees, antennas, flagpoles, or even you. That's why you want to avoid being the tallest object around when lightning threatens. When the two leader strokes connect, the giant flash we know as lightning results. Lightning heats up the air around it to 50,000 degrees, which is hotter than the sun! The superheated air expands quickly, and we hear the resulting shock wave as thunder. If you picture a thrown rock landing in a perfectly calm lake and the little rings that radiate outward from where the rock hit, then you can imagine the shock wave spreading out from a lightning bolt. How far away is the lightning? Simply count the number of seconds between when you see lightning and when you hear its accompanying thunder, then divide that number by five. This is the number of miles away the lightning was.

Between 1996 and 2005, Michigan received a yearly average of 302,614 cloud-to-ground lightning strikes. I have testified in several lawsuits involving lightning, and the resulting injuries (if you even survive) can be significantly life altering. If you are outside and see lightning, remember the 30/30 rule: head inside if less than thirty seconds elapses between lightning and its thunder, and don't return outside until thirty minutes after the last thunder. Although seeking shelter inside a sturdy structure is the safest thing to do when lightning threatens, a car also offers suitable protection as long as you are not touching any metal inside of the vehicle. If you are outside during an electrical storm and you feel your hair stand on end and/or a tingling sensation on your skin, then lightning may be about to strike. If no shelter is nearby, immediately crouch down on the balls of your feet, making yourself as small a target as possible (don't lay on the ground), and cover your ears. Please take lightning seriously: lightning kills an average of 80 people and injures around 300 people every year in the United States.

Is there anything good about lightning? Surprisingly, lightning offers one nice benefit: a lightning bolt changes nitrogen in our atmosphere into a form of nitrogen that plants welcome as a fertilizer when it rains out of the sky. So, a thunderstorm with lots of lightning is great for our lawns and gardens, as I'm sure some of you avid gardeners have noticed in the past.

Hail

Thunderstorms are pretty common, but a sign that a storm has advanced beyond the "ordinary" thunderstorm stage is hail. When a storm's updrafts become strong enough, raindrops are carried high up to a subfreezing area, where they freeze into little pellets. Each trip up and down in the storm accumulates on the pellet another layer of moisture that then freezes. The little balls of ice bounce up and down like ping-pong balls until they become larger and heavier than the updraft can support, then fall to the ground. Thus, hail size is a key indicator of the severity of a thunderstorm: larger hail indicates a stronger updraft, which identifies a stronger storm. In fact, the next time you see large hail, quickly collect some of the largest hailstones (*after* the storm has ended, as falling hail can be very dangerous) and slice one in half with a sharp knife. You'll see the layers—almost like tree rings!

The National Weather Service originally considered a thunderstorm severe if its hail was three-quarters of an inch or greater in diameter, and this threshold stood for many years. However, in 2009 the NWS changed the minimum severe hail size to one inch. I am particularly pleased with this change, because in 1997 I was one of four group leaders on a National Weather Service project to research the criteria on which severe thunderstorm warnings were issued and to offer recommendations on these criteria. The insurance company statistics I acquired were most telling: significant hail damage did not occur with three-quarter-inch hail but increased rapidly starting at one inch. My group recommended the change to one inch, and I strongly endorse this recent move, which will make severe thunderstorm warnings that much more realistic.

Severe Wind

Another indicator that a storm has become severe is damaging winds. Large raindrops in the intense precipitation core of the storm "pull" air down toward the ground. If a sufficient layer of dry air aloft exists, then some evaporation occurs, which cools the air even further. This cooler air is so much denser than the surrounding warm air that it then accelerates downward, pulling strong winds from the upper part of the storm down with it. This accelerating air hits the ground and spreads out much like water would spread out if you poured a pitcher of water onto the ground. The wind that spreads out at the front edge of the storm is strongest, because it is blowing horizontally in the same direction that the storm is moving, and this leading gust of wind that precedes the storm is called a gust front. The National Weather Service identifies a thunderstorm as severe if it has straight-line wind gusts of at least fifty-eight miles per hour. Straight-line winds cause the majority of thunderstorm wind damage and occasionally even reach 100 miles per hour, as we saw in Grosse Pointe Park on July 2, 1997, when a severe storm with wind at 100 miles per hour blew a gazebo into Lake St. Clair. Five of the thirteen people who sought shelter in the gazebo died.

Tornadoes

A more unusual type of thunderstorm is called the supercell, and most tornadoes come from supercells. Once the atmosphere is unstable enough to support powerful thunderstorms with violent updrafts, the next consideration is the wind environment the storm develops in: the storm needs changing wind direction with height (called wind shear) to support supercell characteristics. Here's how it works: Picture a storm with wind blowing from the southeast into the base of the storm. A few thousand feet above the ground, the wind is blowing from the south. The wind is blowing from the southwest by the time you get up to 10,000 feet and, finally, from the west at 15,000 feet. This veering of the wind with height supports air that spirals upward in the storm, eventually causing rotation within the storm.

But what turns that rotation into the violent winds of a tornado? It's another one of those laws of physics, the conservation of angular momentum, and if you have ever seen a figure skater on ice, then you have this one down pat. The law states that if an object of a certain mass is spinning and its diameter gets either larger or smaller, then its speed *must* adjust accordingly. For example, when a figure skater starts her big spin, you will notice her arms (and even one leg) extended out from her body. As she slowly begins to bring her leg and arms in, her speed rapidly increases. If you want to try this yourself, sit in a chair that spins, with arms out to your side, holding a brick or large book in each hand. Get yourself spinning and quickly bring your arms in to your chest: you'll notice an immediate (albeit brief) acceleration in the speed of your spinning.

Now, think about that rotation in the storm. Something causes the vortex to stretch and narrow. As this happens, the law of conservation of angular momentum requires a corresponding increase in the speed of the rotation, and the stronger the wind in the storm is, the stronger the potential winds of the tornado. Most tornadoes are weak, with winds less than 110 miles per hour. However, the rare strongest twisters can produce wind well over 200 miles per hour.

Although most tornadoes come from supercells, some do not. Sometimes, wind interactions between two severe thunderstorms in close proximity to each other cause enough wind shear to spin up tornadoes. Additionally, an approaching line of severe thunderstorms sometimes develops waves or kinks, and the resulting change in low-level wind is enough to generate a tornado. This type of nonsupercell tornado is often referred to as a "gustnado" by meteorologists and is normally weaker than supercell tornadoes and brief in nature.

One other nonsupercell tornado is the fair-weather waterspout, which is most common over the Great Lakes in late summer and fall, when lake temperatures are still relatively warm and when cool, moist air masses begin moving in. Fair-weather waterspouts are generally smaller, briefer, and less violent than supercell tornadoes. However, they still pose a significant threat to boaters nearby and occasionally cause damage if they move ashore.

Fortunately, Doppler radars, which have been used operationally since the 1980s, allow us to see wind inside of developing supercells. Raindrops are targets seen by the radar, and these targets are carried by wind in the storm. The radar sends out a pulse that travels at a specific frequency. When it bounces off raindrops moving toward the radar, the reflected pulse's frequency is shifted a little higher; when it bounces off raindrops moving away from the radar, the reflected pulse's frequency is shifted a little lower. The radar's computer determines the amount of frequency shift and correlates this to a wind speed. So, meteorologists can see both wind speed and direction inside the storm, which gives us the ability to identify strong rotation in developing supercells.

Tornadoes in Michigan

Before discussing Michigan tornadoes in greater detail, it is important to know that the United States receives **95 percent** of the world's tornadoes. "Tornado Alley," as the central United States is sometimes called, is uniquely positioned between the Gulf of Mexico to the south and a large landmass to the north. Warm, moist air surges northward from the gulf, while cool, dry air pushes south from Canada, and where the two collide creates a type of storm not often seen in most other parts of the world. It's like putting matches and gasoline in the same room. You know that something bad will eventually happen.

Most people grossly misunderstand the tornado threat in Michigan. It is true that we don't receive nearly as many tornadoes as Texas and Oklahoma, and because of this, some people think that tornadoes are not a serious threat here. This could not be further from the truth. Tom Grazulis conducted the most comprehensive historical and statistical study of tornadoes in American history, and his publication *Significant Tornadoes* provides tremendous insight into Michigan tornadoes.

For example, the top five states in total number of significant tornadoes received from 1880 through 1989 are Texas, Oklahoma, Kansas, Arkansas, and Iowa (Grazulis defines a "significant" tornado as any tornado rated F2 or greater on the zero-to-five Fujita scale or an F0 or F1 tornado that caused a fatality). Where did Michigan rank? Eighteenth. This is not the whole story, however. Weather patterns shift, and there are times when we are much closer to the bull's-eye. For example, from 1953 through 1989, Michigan ranked *fourth* nationally in tornado deaths. During this same period, Michigan ranked *eighth* nationally in the number of violent (F4–F5) tornadoes as a percentage of all tornadoes received, while Texas ranked twenty-seventh. What does this mean? While Texas received more overall tornadoes than we did here in Michigan during this period, there was a greater chance that any tornado occurring here would be particularly violent. This is significant because although only 1 percent of all tornadoes are violent, two-thirds of all tornado fatalities are caused by this 1 percent of tornadoes.

The greatest tornado threat in Michigan occurs in the southern part of the state, which experiences warm, humid air more frequently than areas farther north. However, *since the modern era of tornado statistics began in 1950, every single county in Michigan has experienced at least one tornado.* It is important that those who live "up north" respect the tornado threat, and history bears this out. On April 3, 1953, an F3 tornado with winds between 136 and 165 miles per hour moved through Manistee, Benzie,

TABLE 1. Tornadoes in Michigan, 1950–2009

COUNTY	1950–2009	2009	COUNTY	1950–2009	2009
Alcona	11	0	Lake	2	0
Alger	6	0	Lapeer	20	0
Allegan	26	1	Leelanau	3	0
Alpena	14	0	Lenawee	31	0
Antrim	9	0	Livingston	24	0
Arenac	7	0	Luce	2	0
Baraga	2	0	Mackinac	5	0
Barry	18	0	Macomb	18	0
Bay	12	0	Manistee	2	0
Benzie	4	0	Marquette	6	0
Berrien	28	0	Mason	5	0
Branch	15	0	Mecosta	9	0
Calhoun	15	0	Menominee	7	0
Cass	14	0	Midland	8	0
Charlevoix	4	0	Missaukee	8	0
Cheboygan	6	0	Monroe	28	0
Chippewa	6	0	Montcalm	11	0
Clare	8	0	Montmorency	6	0
Clinton	17	0	Muskegon	7	0
Crawford	10	0	Newaygo	12	0
Delta	11	0	Oakland	31	0
Dickinson	7	0	Oceana	5	0
Eaton	25	0	Ogemaw	14	0
Emmet	5	0	Ontonagon	2	0
Genesee	41	0	Osceola	16	0
Gladwin	9	0	Oscoda	5	0
Gogebic	3	0	Otsego	3	0
Grand Traverse	4	0	Ottawa	18	0
Gratiot	14	0	Presque Isle	6	0
Hillsdale	23	0	Roscommon	8	0
Houghton	1	0	Saginaw	21	0
Huron	12	0	Sanilac	14	0
Ingham	27	0	Schoolcraft	3	0
Ionia	17	0	Shiawassee	25	0
Iosco	11	0	St. Clair	20	0
Iron	5	0	St. Joseph	9	0
Isabella	13	0	Tuscola	17	0
Jackson	17	0	Van Buren	18	0
Kalamazoo	25	2	Washtenaw	24	0
Kalkaska	7	0	Wayne	28	0
Kent	31	0	Wexford	7	0
Keweenaw	2	0			

Source: Michigan Committee for Severe Weather Awareness and National Weather Service.

Note: A single tornado can cross county lines. Therefore, the sum of the counties will not equal the state totals.

TABLE 2. F5 Tornadoes in Michigan

F5 TORNADO DATE	COUNTIES	KILLED	INJURED
May 25, 1896	Oakland/Lapeer	47	100
June 5, 1905	Tuscola/Sanilac	5	40
June 8, 1953	Genesee/Lapeer	115	844
April 3, 1956	Allegan/Ottawa/ Kent/Montcalm	28	340

Grand Traverse, and Leelanau counties, killing two and injuring twenty-four. Just two months later, an F2 twister struck Iosco County, killing four and injuring thirteen. More recently, on July 4, 1986, an F3 tornado touched down in Wisconsin and traveled into Menominee County, injuring twelve people.

And what about those rare F5 tornadoes glorified in the movie *Twister*? Do those happen here? You bet they do. Michigan has been struck by *four* F5 tornadoes since 1880.

The tornado that hit the Flint suburb of Beecher on June 8, 1953, is one of the most notorious tornadoes in our nation's history. This tornado is the ninth deadliest twister ever in the United States and still reigns as the last single tornado to kill more than 100 people in America (see fig. 3). An F5 tornado will strike again somewhere in Michigan. It's a matter of not if, but when. And contrary to popular belief, tornadoes do not shy away from major metropolitan areas. Just since 1980, a tornado roared right up Main Street in downtown Kalamazoo, and two tornadoes hit the city of Detroit. Making matters worse, urban sprawl is creating bigger "targets" for tornadoes.

Finally, although Michigan tornadoes are most common during the afternoon hours in late spring, summer, and early fall, you need to keep your guard up even during "nontraditional" times of the day or year. On October 18, 2007, eleven tornadoes struck our state (the fifth most for a single day since 1950), and most of them occurred after dark (see fig. 4). Coincidentally, just six years earlier, nine tornadoes touched down in Michigan on October 24, 2001. And what about winter? Although severe weather in the middle of winter is highly unusual here, it occasionally develops if we end up on the warm side of a powerful storm system and if the proper ingredients are in place.

Tornado Safety

The most important thing to remember about tornadoes is that there are things you can do to significantly enhance your chance of surviving a close encounter. Most people hurt or killed by tornadoes are not picked up by the twister and thrown somewhere. Rather, flying debris causes most tornado casualties. Think about it: if an ordinary digital camera becomes airborne in a tornado and strikes you at 135 miles per hour, the camera wins this battle, and you lose. So any tornado safety plan should minimize your exposure to flying debris.

First, take advantage of the advance notice afforded by tornado watches. A tornado watch simply tells you to "watch out." It can be a perfectly sunny day when the watch is issued, but all of the ingredients necessary for severe thunderstorms and tornadoes may come together within the next few hours, and this is the only long-term notice you will have. It is your time to plan and prepare. Make sure your flashlights and radios have working batteries, in the event the power goes out. Also check that everybody with you knows where all of you will go for safety if a tornado threatens. This is also a good time to check on any elderly neighbors and relatives and make sure they are aware of the situation.

A tornado warning is an urgent situation, because this means either that a tornado has been spotted or that Doppler radar has identified a thunderstorm capable of producing a tornado. Tornado warnings are now issued for specific areas in immediate danger. This new process, called storm-based warnings, began October 1, 2007, and issues warnings in polygons based on the storm path, rather than for an entire county. There are many ways you may find out about the warning: sirens, television, commercial radio, NOAA Weather Radio All Hazards broadcasts (provided by the National Oceanic and Atmospheric Administration), the Internet, or even personally observing the approaching storm. Incidentally, the current version of weather radios allow you to program them just to alert you to warnings in *your* county, so you are not bothered by warnings for other areas. You need to take cover without delay when a tornado warning is issued for your area, and I practice what I preach.

Late one summer day, when my oldest son, Jared, was about five months old, I was home alone giving him a bath when my neighborhood tornado siren suddenly went off. I did not go outside to take a look, nor did I check the computer to see what was going on. I immediately wrapped Jared in a towel, grabbed his clothes and my weather radio, and rushed right for the basement. When that siren went off, all I knew was that my county was under a tornado warning, and I had no idea if a tornado was twenty-five miles away or the next street over. I heard on the weather radio that an isolated storm had developed and that Doppler radar detected strong rotation within the storm. The storm was over the opposite end of the county and not moving toward me, so I was not in any danger. The moral to this story is that I remembered what happened to the two women you read about at the beginning of this chapter, and I was not about to take any chances with my son.

If a tornado threatens, here is what you need to know.

1. The general rule of thumb in a building is to go to the basement if it has one and to take cover under a table or in a small room down there. *Stay away from windows.* If you are old enough, you probably remember a time when we were told to open the windows before a tornado strikes. This was later determined to be wrong, because taking the time to open windows not only delays getting to your place of safety but also doesn't help.
2. If you are in a building without a basement, go to a small room in the center of the building's lowest floor. A bathroom is great (as long as it doesn't have windows), because pipes in the walls give you some added protection. A small closet or pantry in the interior of that lowest floor also works well.
3. If you are in a car when a tornado warning is issued for the county you are in and you do not see any particularly threatening weather nearby, then rush to the nearest sturdy building to take cover. If you see a tornado, then you have a decision to make. If

the tornado is distant and traffic allows, drive away from the storm at right angles to the direction the tornado is moving and get to a sturdy building for shelter. If the tornado is close and heading toward you or if traffic is jammed and does not allow an escape, *exit the vehicle immediately* and lay in a low spot, such as a roadside ditch (this helps protect you from flying debris). *Do not take cover on the elevated sides of a freeway under the edge of a bridge.* Even though a very famous video shows a television news crew apparently surviving a tornado under a freeway overpass, the tornado actually passed nearby. If that tornado had moved closer to where they sought shelter, then flying debris may have impaled them there.

4. Like cars, mobile homes offer no protection whatsoever from tornadoes. If you live in a mobile home community, then you need to have a tornado safety plan that involves sturdier shelter than your mobile home. Whether it is a storm cellar or permanent building in your community or a nearby house, store, or other building, you need to pay particular attention after the tornado *watch* is issued, so that you have time to move to sturdier shelter when (or, better yet, before) the tornado *warning* is issued.
5. Boats, obviously, are a very dangerous place to be in severe storms and tornadoes and, at a minimum, serve as lightning rods even in "routine" storms. It is vitally important to check the weather forecast before heading out and to have a weather radio with you to keep you abreast of special marine warnings issued by the National Weather Service. More and more boaters either have radar on board, computers with which they can watch the radar on the Internet, or radar on their cell phones and PDAs. So boaters today can monitor approaching weather like never before. Keep an eye to the sky, and head in at the first sign that a storm is approaching.
6. Finally, my greatest fear as a meteorologist is of a tornado striking a school during the school day. Public schools are supposed to conduct periodic tornado safety drills, but did you know that as recently as 1998, Michigan schools were not required to? I

learned this at a meeting of the Michigan Committee for Severe Weather Awareness (MCSWA), and I was later able to get legislation introduced to amend state law so that these drills would be required. I testified before the state House and Senate education committees about the tornado threat in Michigan and joined Governor John Engler when the "Gross Weather Bill" (get it?) was signed into law. Now, every public school in Michigan is required by law to conduct a *minimum* of two tornado safety drills per year, preferably one just after the beginning of the school year and another in the spring. Go to the MCSWA Web site at **http://mcswa.com** for more information about setting up a tornado safety plan at your school, but there are two vitally important things I will tell you right here. First, have a battery powered NOAA-approved weather radio in the main office, change its batteries twice a year, and *keep the radio turned on in alert mode at all times* (you would be surprised at the number of schools that have their weather radios turned off when I visit). Second, have a way to communicate when the power goes out. Most schools today have intercom systems that go silent if a storm knocks out power. Battery-powered walkie-talkies and megaphones are critical should an emergency occur and you have no power.

Although no tornado safety plan guarantees absolute safety from a direct hit, history proves that the above safety tips greatly increase your chance for survival. Commit these rules to memory.

4 Snow and Ice

If there is a weather discussion topic other than winter that evokes stronger feelings among Michigan residents, then I want to know about it. Winter is either a "love it" or "hate it" season, with little room for middle-of-the-road emotions. Those who love winter subscribe to the philosophy that Michigan is a winter wonderland, with numerous recreational opportunities that the rest of the country south of here simply does not get on a regular basis. Skiing, snowboarding, snowmobiling, and even building a good old-fashioned snowman (or, in the case of my bowling-crazy household, the annual snow bowler) are popular activities, and nothing beats Michigan's scenic countryside after a fresh coating of snow.

Some people, on the other hand, abhor everything associated with winter. More clouds, less daylight, cold weather, driving in snow and ice, and not being able to do their warm season hobbies all contribute to a general season-long misery that lasts until the first tulips bloom. To them, there is nothing positive whatsoever about winter, and global warming is not a problem—it's a solution.

Professionally speaking, winter is a fascinating season, with many forecasting challenges that meteorologists do not face in the warm season. In summer, my forecast involves studying various atmospheric parameters to determine how much rain will fall and whether or not the weather will be severe. In winter, I have to first figure out what type of precipitation will fall, before getting to the rest of the forecast, and as you will read later, that can be quite a challenge.

Winter Storms

The first winterlike storms of the season are usually some of our most intense storms of the entire year. As the jet stream starts dropping back into the states from its typical summertime position up in Canada, cold, Canadian air shifts south over the still relatively warm Great Lakes. As I explained in chapter 3, cold air moving over warmer lower levels creates an unstable atmosphere. In this case, cold air is over the warmer Great Lakes, which also adds moisture to the air. This unstable condition is called the Great Lakes Aggregate, and late-fall low-pressure systems pulling cold air over the Great Lakes become the Godzillas of storm systems. In 2005, four of these intense storms struck Michigan during the first two weeks of November, causing fifteen million dollars in damage, and this is no coincidence: the first two weeks of November, climatologically, are the most likely time for the special combination of cold air and warmer lake temperatures that generate these storms. Historically, two monster storms stand out among the rest.

November 7–12, 1913

The Great Lakes Storm of 1913, called the "White Hurricane" by some, is one of the most intense winter-type storms ever to strike the Great Lakes, and the 235 people who died aboard ships and freighters make this one of the worst maritime disasters in North American history. Mal Sillars and I did extensive research into the meteorology of this storm, which included acquiring as many of the original surface observations as possible and then hand plotting and analyzing surface maps. Our research showed that this is the tale of two storms, with the first storm swinging across the Great Lakes from the northwest on November 7 and

8 (we call these "Alberta clippers," as they swing down from northwest Canada and "clip" the Great Lakes). This first storm is important, because freighter captains trying to make their final runs of the season left port thinking that improving weather behind the storm would persist. They could not possibly know about the second storm, which moved up the Atlantic Coast toward Washington, D.C.; rapidly intensified as Arctic air surged south behind the Alberta clipper; and then curled back toward the Great Lakes. By the time this storm reached Erie, Pennsylvania, at 7:00 p.m. on November 9, its central pressure dropped to an amazing 28.61 inches of mercury.

Lake Huron was hardest hit, with hurricane-force wind blowing continuously for twelve straight hours and with waves over thirty feet battering the ships. The lucky ships were blown ashore—those crews survived. However, thirteen ships, including eight enormous freighters, succumbed to the worst weather in anyone's memory. Most of the ships suffered hull failures and sank. However, once the weather cleared on November 10, one of the big ore carriers, the *Charles S. Price,* was spotted floating *upside down* with no survivors. The marine community was shocked, as no fully loaded Great Lakes ore carrier had ever been flipped over like this.

Here are excerpts from an eyewitness account of the Great Lakes Storm of 1913 written by William Smith, captain of the *Harvester.*

> At three p.m. [November 9] the sea was washing over us so heavy that we had to lessen our speed to prevent the sea [from] breaking in the windows. At this time, the wind was northeast, about sixty miles per hour, and getting cold, about fifteen above zero . . . At nine o'clock it was snowing so heavily that we could not see our stern lights. The roar of the wind resembled a steady roar of thunder. It was impossible to hear the sound of our own whistle . . . At 4:30 a.m. on the 10th, the wind shifted to the N.N.W. the snow decreasing and the wind increasing to about one hundred miles an hour . . . The thermometer was down to zero . . . For an hour and a half or until we were past the island [Michipicaten], it seemed as if we might be lost at any time, as the sea came over us so heavily we could feel the boat settle under the weight of water. At times, a cross sea would roll her over on her side, and in another minute she would be nearly standing on end . . . When we arrived at the Soo,

> 11:30 p.m., we were over a foot deeper than when leaving Ashland [Wisconsin], showing we were coated over with more than 800 tons of ice . . . At any time from six o'clock p.m. the 9th, until five o'clock p.m. the 10th, a mistake in signals or orders would very likely have been fatal, more especially a mistake by the wheelsman.

November 9–10, 1975

The storm of November 1975 is better known as the storm that sank the *Edmund Fitzgerald* than for its meteorology. The *Edmund Fitzgerald* left Superior, Wisconsin, on the afternoon of November 9 and traveled closely with another freighter, the *Arthur M. Anderson*, across northern Lake Superior. Meanwhile, a low-pressure area in central Kansas was heading toward Michigan and would reach Marquette in just twenty-four hours. By 1:00 a.m. on November 10, wind had increased to sixty miles per hour, with accompanying ten-foot waves. The lake became increasingly violent through the day, with twelve- to sixteen-foot waves battering the two freighters. The *Edmund Fitzgerald* started reporting damage to its topside and that it was taking on water and listing. Then, at 4:00 p.m., a wind gust estimated at eighty-five miles per hour struck the ships, knocking out both of the *Edmund Fitzgerald*'s two radars. The last radio communication from the doomed freighter was at 7:10 p.m. The ship went down about seventeen miles northwest of Whitefish Point, with the entire crew of twenty-nine perishing, as the ship sank so quickly that they did not even have time to deploy life rafts.

Early-Season Snowstorms

Early-season snowstorms are more than just depressing to some people and cause for jubilation in others: they can be terribly destructive, due to the fact that foliage has not dropped from trees. Since early-fall snow tends to have higher water content due to milder air temperatures, the snow is heavier when compared to the fluffy powder we get in colder weather. Leaves on trees, which we don't see during the winter, create a large amount of extra surface area on which snow accumulates. When a heavy, wet snow starts piling up on trees with foliage, branches (and sometimes even entire trees) fail under the weight.

Lake-Effect Snow

Another consequence of especially fall and early-winter storms is lake-effect snow. If you have never experienced lake-effect snow, then those who have will say that you have no idea what "real" snow is. Most people in Michigan have seen the occasional storm system drop a foot of snow, but that snow typically falls at an average rate of around an inch per hour over a long period of time. Imagine blinding snow falling at the rate of two, three, four, or even five inches per hour for hours on end! THAT'S "real" snow.

Lake-effect snow results from cold air flowing over the relatively warmer Great Lakes. The greater the temperature difference between the water and the air and the longer the wind's fetch over the lake, the greater the potential is for significant snow. Here's what happens: As Arctic air flows over the warmer lakes, it warms and picks up moisture from the lake surface. The route air travels over water is smooth. However, once that air arrives at the shoreline, it suddenly encounters terrain, trees, and buildings. This added friction slows the air down, but the air behind keeps moving on in. As a result, the moist air (moist due to its trip over the lake) is forced violently aloft, creating lines of intense snow squalls (see fig. 5). Who gets the squalls is highly dependent on the exact wind direction. Even a 5 percent error in the predicted wind direction may mean that a location expecting little or no snow gets inches (or feet) while an area expecting a blizzard gets nothing.

One of Michigan's greatest lake-effect snowstorms occurred December 24–29, 2001, when 2 to 6 feet of snow fell in an area from Traverse City through Petoskey. Petoskey officially received 78 inches, and there were unconfirmed reports of 100 inches elsewhere! Another notable event struck Eben Junction in western Alger County with 35 inches of lake-effect snow February 21–23, 2009.

Forecasting Snow

Whether we get snow from a passing storm system, lake effect, or both, forecasting the amount of snow can run the gamut from easy to extraordinarily difficult. There are so many components to snow forecasting that one hardly knows where to start, so let's begin with the computer models. I already discussed in chapter 1 how meteorologists use computer models to make their forecast. So, the first order of business is to determine which computer model or models has the best handle on the track and intensity of an approaching storm system, as well as the expected temperature and wind before, during, and after the storm.

The heaviest band of synoptic snow (i.e., snow that directly results from the storm itself and not from lake effect) generally falls about 75 to 125 miles to the left of the low pressure center's path. So, determining the path of the low is paramount—but it is not the whole story. Once I decide on a path, I then have to determine if all of the precipitation is going to fall as snow. There is only so much moisture available in the atmosphere for the storm to turn into precipitation (we meteorologists call this *precipitable water*), so if some of that water vapor condenses and falls as rain or ice, then this cuts into the snow totals. After deciding how much of the precipitable water will fall as snow, I then have to take temperature into account. One inch of water generates roughly twelve inches of snow, but that ratio changes based on the temperature. In warmer weather, the same amount of moisture would fall as far less snow, and in colder weather, the ratio is much higher.

There is no better example of the importance of this water-to-snow ratio than to look at February 2003. Unbelievably, Detroit then had its seventh *snowiest* February ever, with 19.2 inches of snow, but also its fourteenth *driest* February ever, with 0.66 inches of total precipitation. How is it possible to have near-record snowiest and driest months in the same month? This occurred because most of the snow fell as fluffy powder with little water content on very cold days, while little or no ice or rain fell on the other days.

Once I determine the storm path, precipitation type, and water-to-snow ratio, there are other considerations to look at, such as upper-air data to identify features that may generate vertical motion so violent as to create thundersnow, which significantly enhances snow intensity. I also look at vertical profiles of the atmosphere to ensure that it is saturated enough so that no large-scale evaporation takes place to cut into my snow totals. Careful analysis of satellite and radar show not only changes in the structure of the storm system but also where any heavier snow bands are setting up. And, of course, I look to see what kind of snow is falling in areas currently affected by the storm. However, changes in the storm's structure as it moves our way change the coverage and intensity of its associated snow, so this is only helpful if it appears that the storm is moving our way in a relatively steady state. Finally, I have to tack on any additional snow accumulation from lake-effect snow before or after the storm itself passes by.

Ice

As you can see, snow forecasting can be quite difficult, but wait until you see what meteorologists have to go through to forecast freezing rain and sleet! Determining our precipitation type involves a very difficult extra step: forecasting the thickness of various temperature layers aloft. Freezing rain occurs when raindrops fall into subfreezing air near the ground and then freeze on contact (see fig. 7). It doesn't matter if they originate as raindrops in the cloud or start as snow and melt on the way down: if a thick enough layer of *above-freezing* temperatures aloft exists, then this water reaches the ground in a liquid state and then freezes. However, increasing the thickness of the *subfreezing* layer of air near the ground causes raindrops to freeze into little balls of ice known as ice pellets, or sleet. Now you see meteorologists' challenge: in addition to forecasting all aspects of a winter storm's path and strength, we also have to predict the lower atmosphere's thermal profile to determine how much snow, ice, or rain will fall, based on the timing of changes in the vertical temperature structure. You also now can further appreciate the importance of the data from those weather balloons I discussed in chapter 1.

If the approaching storm has a significant amount of precipitable water and if the temperature profile aloft is favorable for a sustained period of freezing rain, then there will be significant ice accumulations with devastating results. It seems that we get a memorable ice storm every year, but the ice storm of April 3–4, 2003, will long stand out in people's minds due to both its severity and the number of people affected. This storm, potentially the worst ice storm to hit Michigan in fifty years, deposited between a half inch and one and a half inches of ice across most of the Lower Peninsula. At least 450,000 homes and businesses lost power, and 50,000 of those were without power for a week. Although it is almost impossible to add up the damage cost of a storm like this, the National Weather Service estimates at least 166 million dollars in total damage from this single ice storm.

Ice storms are more common the further south you go in Michigan, because these areas are closer to warm air aloft pushing northward over our colder winter air mass. However, upper Michigan receives its fair share of ice, such as in December 2004: an ice storm hit Chippewa County on December 7, with a quarter to a half inch of ice accumulating from Detour Village through Pickford and Kinross, and another ice storm on December 30 plastered Baraga, Dickinson, Gogebic, Houghton, Iron, Keweenaw, Marquette, and Ontonagon counties with a solid quarter-inch coating of ice. Three days later, another severe ice storm hit Delta, Gogebic, Menominee, and Schoolcraft counties.

Watches, Warnings, and Advisories

Winter Storm Watches are issued when a significant winter storm *might* affect your area within twelve to forty-eight hours. This is a time to plan and prepare.

Winter Weather Advisories are issued in anticipation of three to five inches of snow, less than one-quarter-inch of ice accumulation, blowing snow reducing visibilities, and/or hazardous drifting of snow.

Winter Storm Warnings are issued for expected snowfall of six inches or more in twelve hours or eight inches in twenty-four hours (possibly combined with significant blowing and drifting snow), or greater than one-quarter-inch ice accumulation.

Blizzard Warnings are issued when fierce winds and heavy snow will reduce visibility to near zero and creates wind chills well below zero for at least three hours.

Winter Clouds

Regardless of who gets what kind of winter weather, one thing that all Michiganians get is clouds, clouds, and more clouds—especially in November and December. This adds to our "cabin fever," a condition technically known as seasonal affective disorder, which is chiefly triggered by diminished sunlight. Only the Great Lakes freezing over or mild weather when there is not enough difference between the air and water temperatures negates our lake-effect clouds.

5 Water and Floods

Michigan is home to one of our planet's most remarkable natural resources. The five Great Lakes form the earth's largest surface freshwater system. Combined, their 6 quadrillion gallons of water are one-fifth of the world's freshwater supply and nine-tenths of our nation's supply. In addition, Michigan has numerous inland lakes, rivers, and wetlands that, together, form what are called *watersheds*. Water is a vital part of our state identity.

Anybody who lives near one of the "big" lakes knows that water levels fluctuate over time. Changing water levels are directly tied to precipitation and evaporation, and the more extreme our precipitation or drought conditions are, the more drastic the change in water levels. The replenishing of Great Lakes water is an annual cycle that starts with snow—but not just "any" snow. In the previous chapter, I discussed synoptic snow (snow directly generated by storm systems) and lake-effect snow. A winter storm crossing Michigan brings moisture from somewhere else, such as the Gulf of Mexico, and dumps that moisture on us as snow. Lake-effect snow, by contrast, takes moisture from the Great Lakes and dumps it back down as snow. So, in order to replenish our Great Lakes, we need lots of snow from winter storms, not lake-effect snow. In spring, the winter snowpack melts, and this snowmelt draining into the lakes raises our water levels. While heavy spring or summer rains help, this is normally more than offset by summer evaporation, when our lake levels start lowering. Thus, less synoptic snow over the winter (regardless of how much lake-effect snow we get) results in relatively lower lake levels the rest of the year. I know many people who truly hate snow but love their summer water hobbies. If you fall into this category, you should accept that Michigan needs snow, not just because the winter tourism it generates significantly impacts our economy, but ultimately because snow benefits *you* through its positive impact on our water levels. Without snow, our Great Lakes would shrink.

Flash Floods

Living on the water is a special privilege, but it also can be a curse. Flash floods are a periodic danger but receive less respect than most other types of extreme weather. You should know that, over the long term, flash floods have killed more people in America than any other type of severe weather—more than tornadoes, hurricanes, and winter storms.

There are many reasons that flash floods occur. Sometimes, it is simply a constant barrage of thunderstorms. Not surprisingly, meteorologists call storms crossing the same region one after another "training," as the storms move over the same area like a never-ending line of train cars. Michigan's most famous example of training is the heavy rain event that occurred across the central Lower Peninsula on September 9–12, 1986, one of the worst that this country has ever seen. Persistent heavy thunderstorms dropped unbelievable rain amounts over 14,000 square miles, causing eleven dams to fail, an estimated 400 to 500 million dollars in damage, and the declaration of twenty-two counties as disaster areas. At least ten people died either directly or indirectly from the massive floods, and nearly 100 others suffered injuries (see fig. 6).

Although intensity and duration of precipitation are the greatest factors in deciding whether major flooding will occur, there are other im-

TABLE 3. Heaviest Rain Totals from the Flood of 1986 (in inches)

Big Rapids	19.05	Gladwin	13.64
Saginaw	17.48	St. Johns	13.61
Vestaburg	17.11	Bad Axe	13.39
Evart	17.09	Greenville	13.34
Alma	16.31	Standish	13.11
Essexville	15.86	South Haven	12.65
Mt. Pleasant	15.42	Yale	12.60
Ionia	14.72	Lowell	12.16
Bloomingdale	14.64	Manistee	12.10
Freemont	14.58	Montague	11.84
Marlette	14.52	Hart	11.44
Harbor Beach	14.17	Frankfort	10.99

Note: A 100-year rainfall for *five days* of rain is between six and eight inches.

portant criteria to consider. For example, National Weather Service hydrologists (meteorologists specially trained to forecast flooding potential) take into account how moist or dry the ground is. Dry conditions mean that the ground is able to "soak up" more rain than saturated ground cover, and this means less runoff. Time of year is also important. If the ground is frozen, then it cannot soak up rain or snowmelt, and significant runoff occurs. Another thing to consider is urban sprawl. Buildings, streets, and parking lots cannot absorb water, so replacing natural ground with concrete, brick, and asphalt makes expanding city areas increasingly prone to flooding.

Ice Jams

Another common source of Michigan flooding is spring ice jams. As winter ice melts and breaks up, it starts flowing downstream. Normally, physical contact between the melting ice sheets breaks them up into smaller chunks that continue flowing downstream. On some occasions, however, the ice jams and forms a blockage in the river, resulting in severe flooding upstream. Those who live downstream from a severe ice jam also need to monitor the situation, because rapid breakup of the ice jam can be similar to a dam break, with a corresponding initial rush of high water downstream.

Seiches

The last type of potential flooding has nothing to do with precipitation or snowmelt. Rather, this type of major fluctuation in water level is caused by wind. Anybody who lives along a Great Lake shoreline knows that strong wind can either raise or lower your water level by enormous amounts. This is called a *seiche* (pronounced "say-sh"). The wind required for seiche development can be either strong wind over a long period of time or a shorter period of extremely strong wind.

The Seiche of November 11, 1940

The Armistice Day Storm of 1940 is one of the most notable storms ever to hit the Midwest and is another one of those infamous "Gales of November." This intense storm moved from Colorado toward Minneapolis, which put Lake Michigan on the eastern side of the storm. Strong southerly wind ahead of the storm roared right up the length of the lake, generating enormous waves (some say as high as fifty feet) and a tremendous push of water onto the lake's north shore, causing considerable damage.

The Seiche of October 28–29, 2006

Lake Erie water levels at both Monroe and Buffalo were within a few inches of each other on October 27, 2006. Then, two days of wind blowing from the west at thirty to forty miles per hour generated tremendous seiche conditions. The water level dropped four feet at Monroe and rose four feet at Buffalo. At the peak of the seiche, there was nearly an eight-foot difference in water level between Monroe and Buffalo (see fig. 8)!

The Seiche of May 31, 1998

This seiche resulted from extreme wind ahead of a particularly intense, long-lived line of severe thunderstorms, called a *derecho* (pronounced "deh-RAY-show"). The derecho crossed Lake Michigan between 4:00 and 5:00 a.m., with wind gusts over 100 miles per hour screaming from west to east. The intense wind quickly pushed water levels higher at the

Lake Michigan shore near Muskegon. The tugboat *Stephen M. Asher* was motoring through the channel between White Lake (north of Muskegon) and Lake Michigan when the seiche struck. The boat managed to negotiate the high wind and sudden rise in water level ahead of the storm, but it capsized when an even higher surge of water returned back toward Lake Michigan from White Lake after the wind subsided. Fortunately, the crew managed to safely reach shore.

Flash Flood Safety

The National Weather Service issues flash flood warnings when flash flooding is occurring or is imminent. Remember the following rules when you are in a flash flood situation.

1. Leave areas subject to flooding, including low terrain, dips, valleys, washes, and so on.
2. Avoid already flooded and high-velocity flow areas. Do not attempt to cross flowing streams.
3. If driving, be aware that the roadbed may not be intact under floodwaters. Turn around and go another way. *Never* drive through flooded roadways.
4. If your vehicle stalls, leave it immediately and seek higher ground. Rapidly rising water may engulf the vehicle and sweep it away.
5. Be especially cautious at night, when it is harder to recognize flood dangers.
6. Do not camp or park your vehicle along streams and washes, particularly during threatening weather conditions.
7. Do not let children play around high water, storm drains, rivers, or creeks.
8. If advised to evacuate, do so *immediately*.
9. Move to a safe area before its access is cut off by high water.
10. Monitor NOAA Weather Radio All Hazards, television, and radio broadcasts or the Internet for the latest warnings and information.

6 Heat and Cold

Michigan is perfectly located for both extreme heat and cold, although the Great Lakes do temper some of our extremes and provide us with more relief than some of our neighboring states. This fact was even recognized by H. Pennoyer, a nineteenth-century Michigan settler.

> Our coldest and hardest wind storms, taking their rise in the mountain regions of New Mexico, Colorado and Montana, blow a thousand miles over land with no decrease in the cold of the snow-capped mountains until they reach Lake Michigan. There, in the coldest winters, it stirs up the waters that have retained the heat garnered in from the long summer's sun shining upon the surface of the lake. And as the storms beat upon the east shore they bring with them a warmer and more genial atmosphere than existed on the west shore. As early as 1836 I began to notice the difference between the climate of the west and east shores of Lake Michigan, and to note the difference in favor of Ottawa county and the east shore. The thermometer in this county often stood at zero, at Chicago 18 below the same day and hour, the wind blowing from the southwest.

Fortunately, temperature does not need much explanation. You know when it's hot, and you know when it's cold. But why are some hot days hotter and some cold days colder? As with many other types of weather, we need to start with a look at what is happening aloft, and we'll need a bit more math to explain it.

Back in chapter 1, we looked at the ideal gas law, $PV = nRT$, and we will come back to it in a moment. Right now, though, there's another relationship we have to look at, called the hydrostatic equation. This equation states that $P = g\rho h$, where P is the pressure, g is the acceleration due to gravity, ρ is the density of the air, and h is the height of the column of air. So, this equation states that PRESSURE = GRAVITY multiplied by AIR DENSITY multiplied by HEIGHT. Thus, *pressure* is directly proportional to *height* in the hydrostatic equation and is directly proportional to *temperature* in the ideal gas law. This means that *temperature is also directly proportional to the height of the air column.* To really simplify this, if $P = T$ and if $P = h$, then $T = h$.

Meteorologists use upper-air charts to track rising and falling heights in the atmospheric column. The greater the thickness of the column of air overhead, the warmer it is going to be here at the surface. Conversely, the lower the thickness of the air column above, the colder it is going to be. Thus, areas south of the jet stream tend to be much warmer than areas north of the jet stream, because dips or "valleys" in the jet stream pattern represent falling heights and colder temperatures, while peaks or "mountains" in the pattern represent rising heights and warmer temperatures. Extreme surface temperatures here in Michigan normally result from our being beneath extreme peaks or valleys in the jet stream pattern, and now you know why!

Extreme Heat

Although the Great Lakes tend to moderate extreme air masses crossing the state, we also see occasional periods of extreme heat. Nothing tops the great Heat Wave of 1936, in which most of the state surpassed the century mark not just once, not just twice, but up to *seven* consecutive days (see fig. 9).

Even cities that did not have three or more days over 100 degrees

TABLE 4. The Heat Wave of 1936

CITY	JULY 7	JULY 8	JULY 9	JULY 10	JULY 11	JULY 12	JULY 13	JULY 14
Detroit		104	102	102	101	100	102	104
West Branch		105	103	105	103	105	107	
Alpena		104	102	104	100	105	106	
Traverse City	105	104	102	104	104			
Ironwood					102	103	104	
Munising	103	103	103				100	

still had noteworthy temperatures. For example, Three Rivers hit 107 degrees on July 13 and 14, East Tawas reached 106 on July 8 and 9, Gladwin hit 105 on July 13, and Big Rapids made it to 103 on July 13 and 14. And what was the highest temperature experienced during the Heat Wave of 1936? That dubious honor goes to Mio, which recorded a high of 112 degrees on July 13, still the highest temperature ever in Michigan's history.

There are two factors, humidity and wind, that sometimes make extreme heat feel hotter and extreme cold feel colder. The temperature that it actually "feels like" outside is called *apparent temperature,* which is more commonly called *heat index* in the summer and *wind chill* in the winter (see figs. 11 and 12). Let's explore these two concepts.

Heat Index

Most people know that if we have two 90-degree days, one with low humidity and one with high humidity, we feel hotter on the more humid day. Before understanding why humidity makes us hotter, we have to understand that evaporation makes us colder. You probably notice that you feel cool when you step out of the shower or a pool. This is because water is evaporating from your skin, and energy is needed for the process of evaporation to occur. You feel the energy loss as a drop in temperature. So, humans perspire because evaporating sweat cools us (notice how much cooler you feel when a breeze blows on your wet or sweaty skin, as this hastens evaporation). Yes, perspiration is your built-in air-conditioning system.

On humid days, there is relatively more water vapor in the atmosphere.

Think of the air as a sponge: the wetter the sponge, the less *additional* water it can absorb. So, you feel warmer than the actual air temperature on humid days because less perspiration evaporates from your skin into the air.

When the air is dry (lower humidity), moisture evaporates more quickly from your skin, which promotes a cooler feeling. Ironically, extremely hot days with very low humidity are also dangerous because skin moisture evaporates so quickly that you sometimes do not seem to perspire. You quickly lose fluids but do not realize this because you think you are not sweating. By the time you feel very thirsty, it is almost too late: you are dehydrated. This is why it is so important to drink *before* you are thirsty on hot days. Plain old water is best, followed by sports drinks. *Do not hydrate with alcoholic and caffeinated beverages,* which actually dehydrate you. Some people think that nothing tastes better than a cold beer on a hot day, but excess alcohol (or caffeine) in extreme heat is very dangerous.

The National Weather Service issues a *heat advisory* here in Michigan when the temperature or heat index is expected to be at or above 105 degrees for three or more hours.

Heat Exhaustion

When the body is unable to cool itself due to high heat and humidity (possibly combined with physical activity), then heat exhaustion may occur. Symptoms include headache, dizziness, light-headedness, weakness, irritability, confusion, vomiting, decreased or dark-colored urine, fainting, and pale/clammy skin. If you suspect that somebody has heat exhaustion, move him or her immediately to a cooler (or at least shaded) area, and cool them with a wet cloth and fanning. If they are dizzy or light-headed, lay them on their back and elevate their feet six to eight inches; if they are sick to their stomach, lay them on their side. Loosen and remove any heavy clothing, have them drink a small cup of cool water every fifteen minutes if they are not nauseous, and seek medical attention.

Heat Stroke

If heat exhaustion is not treated, then the person's condition may advance to heat stroke, which is a medical emergency requiring immediate medical intervention. Heat stroke symptoms include dry pale skin that is not sweaty, hot red skin, irritability or confusion, seizures, and loss of consciousness. If you suspect that somebody has heat stroke, call 911 immediately. Proceed as described above with heat exhaustion, and if ice is available, place ice packs in the armpit and groin areas.

Dangers to Children and Pets

Any discussion about heat must also include vehicles parked outside on hot, sunny days. I once had to do a "hot weather story" for my television station, and I kept track of the interior temperature of my news vehicle with a portable infrared temperature sensor. I put all of the windows up and left the car to go interview some people about the hot weather. I returned about fifteen minutes later, and in that time, the interior temperature rose to between 120 and 130 degrees. Even if you leave the windows "cracked" open, the interior temperature will increase 20 degrees or more in just ten minutes. *Do not leave children or pets inside parked cars on hot days.* Call your local police if you see a child or animal locked in a vehicle parked outside on a hot, sunny day.

Hot weather is even more difficult on dogs and cats, which do not sweat like you and me and only get rid of excess heat by panting and through the pads of their feet. If you have pets that stay outside on hot days, make sure they at least have a shady place to escape the sun, and provide lots of cool water.

Extreme Cold

Michigan's coldest temperatures start with a bitterly cold air mass moving south from Canada, and our coldest temperatures occur on nights with clear skies and light wind, which allow for efficient *radiational cooling*. Whatever heat we get during the day radiates upward once the sun sets, but on cloudy nights, some of that heat is absorbed by clouds and reradiated downward, which generates a bit of warming. If the night is breezy, then the heat radiating up is mixed throughout the lower atmosphere, which cannot cool as much as it otherwise could. On clear, calm nights, heat radiating to space is neither inhibited by clouds nor mixed

TABLE 5. February 17, 1979

CITY	LOW TEMPERATURE
Trout Lake	–43
Herman	–40
Detour Village	–39
Alpena	–37
Gaylord	–37
Pellston	–37
Traverse City	–37
Sault Ste. Marie	–35
Houghton Lake	–34
West Branch	–28
East Tawas	–26
Maple City	–24
Standish	–24
Harrisville	–20

by wind, and temperatures drop nearly all the way to the dewpoint temperature. Since drier air has lower dewpoint temperatures, you can now see that cold, *dry* air has the greatest potential for large drops in nighttime temperatures. Given that the temperature can never drop below the dewpoint temperature, meteorologists frequently use the expected nighttime dewpoint temperature as a starting point when forecasting low temperatures on good radiational cooling nights. Finally, since colder air is denser (i.e., "heavier"), it tends to settle into valleys and low spots, so these locations are typically colder than surrounding areas on radiational cooling nights.

So how cold can it get in Michigan? The coldest temperature ever in our state was an unbelievable –51 degrees in Vanderbilt on February 9, 1934. Temperatures colder than –40 degrees are infrequent, but they do occur. One particularly nasty example is three straight days with record low temperatures of –45, –44, and –41 degrees recorded by a cooperative weather observer eight miles west-northwest of Stephenson on February 3–5, 1996. One of Michigan's coldest days ever (judged by the widespread top ten coldest temperatures ever) occurred on February 17, 1979.

Identifying one of Michigan's coldest cold waves in the western Lower Peninsula requires a history book: it occurred in February 1899.

TABLE 6. **February 1899**

CITY	FEBRUARY 10	FEBRUARY 11	FEBRUARY 12	FEBRUARY 13
Baldwin	–36	–49	–48	–37
Big Rapids	–33	–36	–34	
Hastings		–26	–31	–24
Muskegon		–30	–29	–22
Grand Rapids	–21	–21	–23	–24

Even southeast Michigan sometimes experiences extreme cold, with many locations experiencing all-time record low temperatures between –20 and –30 degrees.

Wind Chill

Extreme cold is bad enough, but wind makes it even worse because it increases the rate of heat loss on exposed skin. Wind not only dissipates the layer of warmer air near our bodies; it also hastens evaporation of moisture from our skin (remember that evaporation causes cooling). Keep in mind that wind chill is a perceived temperature: it only affects living things and does not affect your car, mailbox, sidewalk, or driveway.

The first attempts to quantify various combinations of wind and temperature and how much colder they make you feel occurred in 1945, when two Antarctic explorers, U.S. Army major Paul Siple and geographer Charles Passel, measured the cooling rate of water in a container hanging outside. Their wind chill chart became the gold standard for over five decades, until research in the 1990s found that its chill values were too severe. Subsequent research using human beings resulted in a more accurate wind chill formula. In these tests, six men and six women were exposed to different temperatures and wind speeds in a wind tunnel. Each person had thermal transducers affixed to various spots on their faces to measure heat flow, while they walked at three miles per hour on treadmills in the wind tunnel. Based on this research, the National Weather Service released its new wind chill temperature index on November 1, 2001, which also includes exposure times at which frostbite occurs on exposed skin. Although the criteria vary slightly across the state, in general the National Weather Service issues a *wind chill advisory* when wind chills are expected to be between –15 and –24 for southern Michigan, or between –20 and –29 for northern and upper Michigan. A *wind chill warning* is issued for expected wind chills of –25 or colder in southern Michigan and –30 or colder in northern and upper Michigan.

Dressing for Cold

Dressing in layers of loose-fitting, lightweight warm clothing really works. Air is a tremendous insulator, and every layer of clothing you add also adds a layer of insulating air. The outer layer should be something that repels water and moisture to keep you dry. The next most important thing is to wear a hat and mittens. We lose 30 to 50 percent of our body heat through the head, so a hat helps mitigate this important source of heat loss. Mittens are a better option than gloves, so wear mittens if you have a choice. Also, cover your mouth with a scarf to help protect your lungs in extremely cold weather.

Frostbite

Frostbite occurs when tissue in the skin freezes, and it most commonly develops on the extremities—hands, feet, ears, and nose. Frostbitten skin may become numb and will look pale or white. Treat the area by warming it *slowly* (do not rub it), and see a doctor immediately.

Hypothermia

Hypothermia occurs when the body temperature drops below a safe level (generally, under 95 degrees). Warning signs include uncontrollable shivering, memory loss, disorientation and incoherence, slurred speech, drowsiness, and apparent exhaustion. If you suspect that somebody has hypothermia, seek immediate medical attention, then start warming the body *slowly*.

Every winter we hear about people dying from overexerting themselves in cold weather. If you are elderly, have a heart condition, or have high blood pressure, try to limit strenuous physical outdoor activity during winter. Cold weather is known to increase blood viscosity, narrow small blood vessels, and increase blood pressure, all of which puts added strain on the heart. Do not let yourself become a statistic: use common sense in winter weather.

7 Global Warming

Without question, global warming was this book's most difficult subject to write about, but the difficulties do not arise in explaining the science. Rather, my challenge lies in dealing with your perceptions (correct and incorrect) about global warming. What started as a purely scientific issue immediately transitioned into a political issue when policy decisions started being discussed. I will only present accepted scientific facts about our changing climate and examine the potential impact of those changes on Michigan. *I will not champion any political point of view, and everything written in this chapter is written from my vantage point as a scientist and without political consideration whatsoever.* Of course, new research is being published all the time, so everything discussed here should be considered current through February 2010.

The Greenhouse Effect

Any discussion about global warming must begin with an explanation of the greenhouse effect, which has taken on a significantly negative connotation in recent years. In reality, without the greenhouse effect, Earth would be so cold that life as we know it would not exist.

A planet's average temperature is determined by how much incoming solar radiation the planet receives and how much of that energy the planet radiates back to space. One of the factors that determine just how much solar radiation a planet absorbs is the color of the planet's surface, and perhaps you already know that darker colors absorb more radiation than lighter colors. For example, I once measured the temperature of a concrete sidewalk on a hot, sunny day and compared it to the newly seal-coated asphalt driveway right next to it. The darker asphalt was around 25 degrees hotter than the concrete. This same effect is why light-colored clothing keeps you cooler on a hot, sunny day than dark clothes.

Another factor that impacts how much solar radiation a planet retains is the composition of its atmosphere. These atmospheric components are actually transparent to visible light but absorb radiation emitted both from the sun and the planet and reradiate that radiation back down to the surface. Here on Earth, the most important radiation-absorbing gases in our atmosphere are water vapor, carbon dioxide, methane, and nitrous oxide. These gases are called *greenhouse gases,* and the process by which our atmosphere absorbs radiation and sends it back down to the surface is called the *greenhouse effect.*

Earth's two planetary neighbors are perfect examples of greenhouse extremes. A thick atmosphere of carbon dioxide enshrouds the planet Venus. Solar radiation penetrates the atmosphere and reaches the surface, but the carbon dioxide absorbs and radiates so much energy back down to the surface that the planet's average temperature is a blistering 900 degrees (even hotter than Mercury's surface temperature).

By contrast, Mars has a thin atmosphere of carbon dioxide. Abundant sunlight reaches the Martian surface, but much less radiation is absorbed by its atmosphere and reradiated back down to the surface. As a result, Mars' average surface temperature is –81 degrees.

I think you now realize just how special Earth is: the combination of our distance from the sun and just the right amount of greenhouse gases in our atmosphere results in perfect temperatures for life as we know it to develop and flourish. With any more or less of these greenhouse gases, we very well could have resembled Venus, Mars, or even the

Moon if we had no atmosphere at all. At this point, you now also appreciate the primary concern of climate scientists worldwide: human activity has changed the composition of our atmosphere by increasing the proportion of greenhouse gases through emissions from transportation, power generation, and industry.

In August 2007, I had the privilege of serving on a panel in a national teleconference with three of the world's leading climate scientists: Dr. Gerald Meehl (senior scientist in the Climate and Global Dynamics Section at the National Center for Atmospheric Research), Dr. Richard Somerville (distinguished professor emeritus from Scripps Institution of Oceanography at the University of California at San Diego), and Dr. Kevin Trenberth (head of the Climate Analysis Section at the National Center for Atmospheric Research). All three of these scientists were coordinating lead authors of the 2007 report on global warming by the Intergovernmental Panel on Climate Change (IPCC). This report, issued once every five or six years by a division of the United Nations, is the foremost authoritative analysis of global warming. There is a small, but very vocal, minority that attacks the validity of the IPCC report for various reasons, calling it a political document. This could not be further from the truth.

I specifically asked Dr. Trenberth about the team of international scientists involved in the IPCC process. He replied that there were over 450 lead authors, over 800 contributing authors, and over 2,500 reviewers from over 130 countries and that the authors received over 30,000 comments (every one of which was documented and addressed) just for part 1 of the four-part report. Dr. Meehl added that the process included extremely vigorous disagreements and debate and that the IPCC report represents not necessarily a consensus of personal opinions of the scientists involved but, more important, *their collective judgment on the balance of scientific evidence.* Dr. Somerville emphasized that politics had nothing to do with the process whatsoever: this process, he said, represents the highest standards in science, with no political influence at all.

There is no question that the IPCC report represents a joint effort of a majority of the world's most eminent climate scientists, who intensely studied every bit of available research to arrive at their conclusions. If there were unanswered scientific questions with varying views among the scientists, then this was openly addressed in the report. So, let's delve into some of the most important questions about global warming.

Is Global Warming Really Happening?

The answer is yes, our planet is currently experiencing a significantly abnormal rate of warming that is very likely outside the realm of any natural variability. Using ice cores drilled in Antarctica, paleoclimatologists have created a time line of global temperatures and greenhouse gases going back 800,000 years. As you would expect, there was quite a bit of natural variability over those years (both significant warming and cooling), and there were even some periods where Earth was warmer than today. But these warm periods resulted from changes in Earth's orbit and ocean circulations. By contrast, our planet's climate over the past 10,000 years has been comparatively stable until the last 100 years, when warming began. Using tree ring data and other "proxy climate data," scientists have a bit more detail about global temperatures over the past 1,000 to 2,000 years. Our current *rate of warming* is greater than that of any warm period in the entire 800,000-year paleoclimatic record and cannot be solely explained by things like El Niño and so on. As a result of numerous studies using various paleoclimatic methods, there is nearly unanimous scientific agreement about Earth's current warming status. So, yes, our planet is getting warmer, at a highly unusual rate of warming.

What Is Causing Global Warming?

Scientists have quantified the amount of greenhouse gases in our atmosphere over the past 800,000 years by studying trapped air buried in deep layers of Antarctic ice. Based on this ice core data, atmospheric levels of carbon dioxide and methane are currently at the highest levels seen in the entire 800,000-year record. Specifically, Earth's current atmospheric carbon dioxide concentration has now surpassed 383 parts per million, compared to a range of 180 to 300 parts per million over the past 800,000 years. Just before this book went to press, another independent research project concluded that current carbon dioxide levels are the highest they have been in the past 2.1 million years.

In addition, while the carbon dioxide rate of increase never exceeded thirty parts per million per *one thousand* years, atmospheric carbon dioxide has increased thirty parts per million just in the past *seventeen* years. Some greenhouse gases are created solely by human activity, while others also occur naturally, but scientists say that their rapid increase in our atmosphere over the past 250 years is not the result of natural causes. In short, human activity is significantly altering the composition of Earth's atmosphere (see fig. 13).

Scientists also created an 800,000-year temperature time line by studying changes in deuterium (a hydrogen isotope with twice the mass of ordinary hydrogen, also called "heavy hydrogen") in the ice cores. When the temperature data was compared with the greenhouse gas time line, **every single interglacial warm period over the past 800,000 years corresponded exactly with a significant increase of atmospheric greenhouse gases** (see fig. 14). In each of those instances, research indicates that changes in Earth's orbit initiated moderate warming, which then caused a massive release of greenhouse gases into the atmosphere, resulting in skyrocketing global temperatures. It is important to remember that these previous warm periods *started* with changes in Earth's orbit, which has not occurred with our present warming.

Some people challenge the current rate of warming, claiming that increased urbanization has artificially skewed temperatures upward. The IPCC specifically researched this issue and determined that temperatures averaged over hemispheric or continental scales were affected by urbanization less than 0.006 degrees Celsius per decade. Furthermore, significant warming has also been measured in *ocean* surface temperatures, where there is no urban impact. Thus, any warming effects from urbanization are negligible from a global perspective.

Earth has most certainly been this warm before, and scientists are confident they know why it got that warm. There are obvious natural impacts on Earth's climate (for example, solar output changes may have impacted some of the early twentieth-century warming), and the IPCC authors made it perfectly clear that they examined all possible causes. One key area of research is the development of computer models that project changes in Earth's climate. Scientists assess the reliability of the models' ability to predict long-term climate by first seeing how they "predict" our current warming. *To date, there is not one climate model that successfully replicates the extraordinary warming of the past fifty years without the addition of human-created greenhouse gases.* In fact, climate models that only consider natural causes show that cooling should have occurred over the past few decades, and this clearly has not happened.

Climate scientists are very confident that increasing greenhouse gases in our atmosphere is the cause for most of Earth's rapid warming over the past half century and that anthropogenic (i.e., human) activity is the primary reason for the increase in greenhouse gases.

What Is Global Warming Doing to Earth?

Earth has already responded to the rapid warming. Here are some of the global effects.

1. Average global temperatures have increased, especially since 1950. Average temperatures in the Arctic have risen twice as fast as the rest of the world over the past 100 years.
2. There has been a likely increase in the number of heat waves and a decrease in the number of frost days and record cold days in midlatitude locations. In fact, record high temperatures in the United States now occur twice as often as record lows.
3. Globally averaged ocean temperatures have increased, especially since 1955.
4. Mountain glaciers, Arctic ice, the West Antarctic ice sheet, and the Greenland ice sheet are diminishing as ice melts due to increased air and ocean temperatures. For example, only 26 of the 150 glaciers counted in Glacier National Park in 1850 still remain. The melting freshwater ice is changing the salinity of the oceans, which may eventually impact important ocean circulation patterns.

5. Globally averaged sea levels have increased during the twentieth century, with the most rapid increase occurring since 1993. Sea levels are now rising at the rate of 0.13 inch per year.
6. Ocean acidity is increasing, as oceans absorb increased carbon dioxide from the atmosphere. As a result, shellfish, which have carbonate shells or skeletons, may suffer significant declines over the next fifty years.

Most people have no idea just how complicated the global warming scenario really is. As the planet warms, consequences cause other consequences, like falling dominos, and the key goal for climate scientists is to determine the importance of each particular falling domino. For example, as polar ice diminishes, it is replaced by water. The darker water more readily absorbs incoming solar radiation, whereas the highly reflective ice reflects much of that radiation. The change from ice to water enhances polar warming and creates a situation that further accelerates ice reduction. In other words, global warming reduces polar ice, which causes more polar ice to melt even faster. Computer models project an initial steady loss of Arctic ice, then an abrupt decrease of this ice occurring sometime between 2020 and 2030. However, scientists were stunned by a sudden ice loss in 2007 that shattered the 2005 record low, followed by a 2008 ice minimum that stopped barely short of the 2007 record, and now by a 2009 ice minimum that was the third lowest since statellite measurements began. They now fear that the abrupt change is occurring much earlier than anticipated. To put this ice loss in perspective, satellite measurements considered along with ship and aircraft observations indicate that as much as *50 percent* of the Arctic ice that existed in the 1950s is now gone, with the rate of ice loss now at 10 percent per decade.

Here is another example of this domino effect: Many high-latitude locations have what is called permafrost, or permanently frozen ground. A considerable amount of organic material lies beneath the permafrost. As global temperatures rise, the permafrost begins to thaw, which allows this organic material to start decomposing. Some of this decomposing material emits carbon dioxide or methane as a by-product. So, global warming causes melting permafrost, which causes organic material below the permafrost to decay, which releases additional greenhouse gases into the atmosphere, which may further accelerate global warming. It's sort of like compounded interest, except that we are compounding temperatures instead of dollars.

There are also many uncertainties, such as the impact of clouds and aerosols on long-term climate, how ocean circulation patterns will change, how much global warming will impact the frequency and intensity of El Niño and La Niña, and so on. Climate scientists still do not have a complete understanding of all aspects of the climate system and their complex interactions with each other, but climate computer models are improving and continue to shed new light on exactly what global warming will do to our planet.

What Will Global Warming Do to Michigan?

Global warming will affect Michigan in many ways, both directly and indirectly. One direct influence recently uncovered is the migration of nine species of Upper Peninsula mammals studied by a group of researchers from the University of Michigan, Michigan State University, and Miami (Ohio) University. Of the nine species, four showed increased population over the past thirty years, while the other five decreased. Most significant is that ALL four increasing species were southern species, while ALL five decreasing species were northern species. After ruling out forest regeneration and human influence, the reason for the migration became clear: as the climate warmed, southern species moved north and replaced some northern species.

Most of global warming's effects will negatively impact nature and society, but there will be some benefits. Here is what scientists project.

1. Michigan may see shorter winters, with fewer severe cold outbreaks. While this would lower heating costs, keep in mind that periodic harsh winter temperatures help keep certain pests in check. For example, milder winters have allowed winter ticks to flourish on Isle Royale, which are infecting more and more moose,

the primary food source for the island's wolves. The moose population has dropped by 65 percent since 2002, and this diminished food source will decimate the wolf pack. In another example, Purdue University scientists released a study in December 2008 stating that milder winters may allow increased winter survival of four different types of pests that attack corn crops.

2. Shorter winters also mean diminished opportunities for cold-weather recreational activities but a potentially longer season for warm-weather sports and hobbies.
3. Shorter winters create potentially longer growing seasons. However, keep in mind that blossoms blooming earlier in spring would also be susceptible to killing frosts, as we saw in 2002 when record warm temperatures brought out cherry blossoms, only to have the crop virtually destroyed one week later by a killing frost, resulting in Michigan's worst cherry crop ever.
4. Michigan may experience less winter ice, and we are already seeing signs of this. Michigan State University Extension horticulturist James Nugent studied freezing trends in Grand Traverse Bay from 1851 to the present and has identified a significant decline of bay freezing and ice duration over the past twenty-five to thirty-five years. Less ice means more open water, which means increased wintertime evaporation, and this contributes to lower lake levels.
5. Michigan may receive more extreme heat, although nighttime temperatures are expected to increase more than daytime temperatures (some scientists say that our summer will feel more like a Missouri or Oklahoma summer by the end of the century). Even the most conservative climate models suggest that Detroit will experience an average of four to eight 100-degree days per summer by the end of the century. Extreme heat stresses humans: Dr. Karin Schenck-Gustafsson, a cardiologist at Sweden's Karolinska Institute, said at the European Society of Cardiology's 2007 annual meeting that a majority of the 35,000 people who died due to the severe European heat wave of 2003 were elderly people with heart problems exacerbated by heat stress. Extreme heat also negatively affects crops and livestock.

6. Winter and spring precipitation may increase 20 to 30 percent, with more of this falling as rain. Heavy rain events may increase, with flooding that is now considered "routine" increasing in severity. Total summer precipitation may decrease by 10 percent, which, combined with hotter summer temperatures, would have a negative impact on our state's vital agricultural sector.
7. Less wintertime ice, higher summer temperatures, and less overall precipitation would lower our lake levels. Even an increase in precipitation would be more than offset by increased evaporation due to higher temperatures. Shallower lakes and rivers will be warmer in the summer, thus affecting many fish species. Cold- and cool-water fish (trout, whitefish, northern pike, and walleye) may decline in southern areas, while warmer-water fish (smallmouth bass and bluegill) may expand northward.
8. Lowering lake levels would also drastically affect Michigan's coastal wetlands, which a large majority of Great Lakes fish depend on for reproduction. Farther inland, smaller streams and tributaries could dry up if not sufficiently replenished, thus negatively impacting associated wetlands and their habitat for wildlife. At a minimum, decreased water volume and stream flow would negatively impact water quality.
9. Changing temperature and precipitation patterns may radically alter the composition of Michigan's magnificent boreal forests, as conifers (i.e., pines) are potentially replaced by trees more acclimated to a warmer climate. There is still some uncertainty as to how drastic the change will be, but future generations may experience a different type of forest on their travels across Michigan. Tree death rates in old-growth forests of the western United States have already more than doubled over the last few decades, and scientists working for the United States Geological Survey (USGS) say that regional warming is the primary cause.

The bottom line is that, yes, Earth has been this warm before, but this time, the warming is occurring in a very, very short period of time (possibly in less than 100 years) due to humans changing the composition of our planet's atmosphere. Some species may be able to adapt, but some will not. In 2007, the Department of the Interior asked the USGS to research the impact of Arctic ice reduction on polar bears. The polar bear, which specifically evolved as a species that uses sea ice as a platform to hunt seals, simply will not have time to adapt. As such, the USGS predicts that rapid reduction of sea ice will cause an estimated two-thirds reduction in the world's polar bear population within fifty years. But has anything like this happened in the past? The answer is yes. A group of scientists studied the fossil record of many different types of mammals, from rodents to giraffes, in what is now northern Pakistan to see how they responded to a dramatic shift in climate eight million years ago. That climate change caused a significant change in vegetation, and most of the species studied became locally extinct, instead of adapting to the new ecosystem.

Humans will adapt, but it will not be easy for everybody as a warmer global climate dramatically impacts some nations. If the central United States becomes hotter and drier, then our nation's agricultural heartland may shift northward. If projections for sea levels are correct, then island nations and coastal locations should be very, very concerned (Dade County in Florida has already created the Miami-Dade Climate Change Advisory Task Force to plan and prepare for anticipated changes due to global warming). If ocean temperatures continue increasing, then hurricanes that develop in favorable atmospheric conditions could be stronger. In fact, research published in February 2010 indicates that while the globally averaged frequency of tropical cyclones may *decrease* by 6–34 percent, there will be a corresponding increase in stronger storms, with globally averaged tropical cyclone intensity *increasing* 2–11 percent, along with rainfall rate *increasing* 20 percent in the most intense part of the storm. Thus, a warmer world may mean fewer, but stronger, hurricanes.

There is still much to determine about the impacts of global warming, but as much as politics have clouded the issue, it is important to remember that *science*—the balance of scientific evidence—indicates that our current rapid rise in global temperatures is largely the result of human activity. A 2008 Harris poll of 489 American climate scientists who were members of either the American Meteorological Society or the

American Geophysical Union indicates that *84 percent* of these scientists agree that human-induced warming is occurring, 11 percent are unsure, and only 5 percent believe that human activity does not contribute to greenhouse warming. Much more uncertainty arose when the questions turned to future effects of the warming, but the overwhelming majority of climate scientists (compared to less than half in 1991) now agree that a significant increase in human-generated greenhouse gases has caused our planet to warm.

"Climategate"

Late in 2009, somebody broke into an e-mail server at the United Kingdom's University of East Anglia's Climatic Research Unit (CRU), stole 3,000 personal e-mail communications from climate scientists and then leaked a few of these e-mails to the press. The decided minority who oppose the 2007 IPCC report (most of them not scientists) then initiated a massive public relations campaign trumpeting these e-mails as "proof" of a "conspiracy" among climate scientists to hide and manipulate data to support their conclusions. This could not be further from the truth. Here's what you need to know.

1. E-mail allows us scientists to rapidly communicate with each other to informally discuss and debate the latest research. However, the illegally obtained e-mails are only personal communication and do NOT represent the final, peer-reviewed, debated science that appears in scientific journals (which forms the basis for the IPCC conclusions). For example, let's say you e-mailed a friend that you wanted to buy a new car, but said that you didn't have the money for it. What if somebody then stole that e-mail, and used it to publicize that you were having severe money problems? Would that be fair? Of course not. Your e-mail was stolen, taken out of context, and the person who stole it had no factual information about the specifics of your financial situation. This is exactly what happened to the CRU scientists.

2. There is an overwhelming amount of scientific research conducted independently by climate scientists around the world that strongly supports the IPCC conclusions. I was in touch with many of these scientists immediately after the "climategate" story broke, and every single one of them confidently said that the stolen e-mails will not in any way alter their research conclusions, which have been independently verified by colleagues at other research institutions and in other nations and have not been proven wrong in numerous attempts by global warming skeptics. To believe that a single scientist can manipulate the world's entire scientific community on global warming is as ridiculous as believing that smoking is good for your health.

Yes, "climategate" is a public relations disaster that will only serve to confuse the public. However, we scientists are not concerned about climate change *science* itself, which is rock-solid and has withstood *scientific* scrutiny for many years now.

Is There Anything We Can Do to Help Reduce Greenhouse Gases?

Believe it or not, there are many simple things you can do that make a significant impact, and most of these will save you money, too. In general, anything that reduces the amount of energy you use is a huge plus, because the less energy you use, the less energy your power company has to produce, which reduces greenhouse emissions. Here are some excellent tips from the Midwest Energy Efficiency Alliance.

1. If you use air-conditioning in the summer, dial your thermostat up a few degrees. Each degree you dial up above 72 degrees reduces your cooling bill by approximately 3 percent. Also, don't let the sun heat up your home. Keep blinds or drapes on sun-exposed windows closed in the daytime to block out solar heat, then open them at night to allow heat buildup to escape.
2. In winter, turn your thermostat down to 68 degrees during the day. For every degree you lower your heat in the 60- to 70-degree

range, you'll save up to 5 percent on heating costs. If you leave home for an extended period of time, lower your thermostat to 55 degrees, which saves you 5 to 20 percent on heating costs. Also, let the sun help heat your home by leaving blinds or drapes on sun-exposed windows open in the daytime, then close them at night to help hold the heat in.

3. Install a programmable thermostat to automatically lower or raise the temperature during regularly scheduled times you are home and away.
4. Replace or clean your furnace filters once a month, and get your furnace tuned up. Proper furnace maintenance can save up to 5 percent on energy costs.
5. Check for drafts around walls, floors, the roof, windows, and doors. Seal leaks between moving parts (a door and frame) with weather stripping. Fill leaks between nonmoving parts (a window frame and wall) with caulking.
6. A properly insulated home can save you 30 percent on winter heating and summer cooling costs.
7. Set your water heater to the "normal" setting or 120 degrees Fahrenheit, unless the owner's manual for your dishwasher requires a higher setting. Wrap your water heater with jacket insulation, following the recommendations of the manufacturer. Savings from these two steps can be 7 to 11 percent of water heating costs.
8. Unplug electronics, such as TVs, VCRs, DVD players, game consoles, computers, monitors, scanners, printers, and appliances, when not in use (and especially when you go on a trip), or put them on a switchable power strip that you can turn off, because they consume power even when not being used. Unplug the *chargers* for your cell phone, PDA, laptop computer, and other electronic devices when not in use, as they continue to use electricity whenever they are plugged in.
9. Replace older, inefficient heating, ventilating, and air-conditioning (HVAC) equipment with ENERGY STAR products, which earn 90 percent or higher efficiency ratings. When replacing other appliances, such as your clothes washer, refrigerator, and

other home appliances, look for the ENERGY STAR logo, which guarantees that the appliance you buy is at least 15 percent more efficient than an unrated version. When you do use these appliances, wait until you have a full load before running them.

10. Replace your incandescent lightbulbs with compact fluorescent light (CFL) bulbs whenever possible. Keep in mind that some CFL bulbs cannot be used on lights or fixtures with dimmer switches, so make sure you buy the right bulb. CFL bulbs use 75 percent less energy than standard incandescent bulbs while providing the same amount of light. For example, if you replace your four most used 100-watt incandescent bulbs with four comparable 23-watt compact fluorescent bulbs, you save about $108 over three years by getting 400 "watts" of total light from bulbs that use only 92 watts of energy. If all U.S. households did this, we could save as much energy as is produced by 30 power plants annually (and remember that these bulbs last much longer than "regular" lightbulbs).

    Very important: CFL bulbs contain a small amount of mercury and should not be thrown in the trash like normal lightbulbs; dispose of them as you would normal household "hazardous waste," and check with your local community about how and where to do this. Also, special steps need to be taken in the event you break one of these bulbs in your home. The Environmental Protection Agency has excellent information at http://www.epa.gov/mercury/spills/index.htm#fluorescent. Do not let these cautions scare you away from CFL bulbs, as the benefits and significant money savings over the life of the bulbs far outweigh the negatives. I have replaced many of the incandescent bulbs in my home with CFL bulbs, and you should too.

11. Replace traditional Christmas lights with light-emitting diode (LED) Christmas lights, which use between twelve and thirty-five times less energy. LEDs also last twenty-five to fifty times longer than the traditional bulbs.
12. Turn off lights when you are not using them. I am amazed at the amount of wasted electricity I see from lights left on for hours in

unoccupied rooms in homes and office buildings. If you leave outdoor floodlights on at night for security, consider installing motion detector floodlights. I did this around my house, and my floodlights only come on at night when a person or car comes up my driveway or when somebody walks behind my house. This is a great energy saver.

Another thing you can do is drive fuel-efficient vehicles and drive as fuel-efficiently as possible: combine errands, carpool, and keep your vehicle tuned up. The less gasoline we burn, the less greenhouse gases we put into the atmosphere. Local governments should also make synchronizing traffic lights a high priority to keep traffic flowing. In fact, my community has replaced some traffic lights with roundabouts; doing this significantly minimizes the amount of time cars spend idling at red lights. Finally, think about planting some trees. Trees take carbon dioxide out of the atmosphere and, when planted in the right place, also provide shade for your home and help reduce summer cooling costs.

Global warming is like a runaway freight train: we are at the point where we cannot stop it, but we are in a position where we can slow it down. A National Academy of Sciences report issued in January 2009 states that if carbon dioxide emissions this century continue increasing at the current rate, some of the atmospheric and oceanic changes we are now seeing will become irreversible for at least 1,000 years after reduction of the emissions. Our climate is changing so fast that some species will not have time to adapt. By taking some or all of the above steps and by also educating our children (as we do with seatbelt use and the dangers of smoking, drinking, and drug use), perhaps we can slow down the rate of warming and give some species enough time to adjust to the changes.

8 Miscellaneous Questions

Television meteorologists receive questions about everything under the sun, and I have answered more than my share over the years. While some of them do not fall under the category of extreme Michigan weather (or even weather), no book about Michigan weather would be complete without at least briefly answering the following questions.

Do Hurricanes Ever Hit Michigan?

No, hurricanes do not hit Michigan. Hurricanes only form over tropical ocean waters at least 82 degrees or warmer, and they quickly diminish upon hitting land, when they move off of their energy source. However, the *remnants* of hurricanes sometimes track all the way to Michigan.

Bill Deedler, weather historian at the Detroit/Pontiac National Weather Service Weather Forecast Office, has actually researched this topic. He determined that roughly two hurricane remnants per decade affect the Great Lakes region, and one biggie stands out among the rest: the September 1941 hurricane (hurricanes were not officially given names in those days). This storm made landfall near

TABLE 7. Rainfall, September 12–14, 2008 (in inches)

Frankenmuth	6.62	Hell	5.58
Grand Blanc and Linden	6.16	Tipton	5.40
Ortonville	6.06	West Bloomfield	5.37
Burt	5.85	Farmington Hills	5.36
Richmond	5.77	Salem	5.35
Chelsea and Howell	5.70	Corunna and Durand	5.34
Highland and Livonia	5.70	White Lake	5.29
Clarkston	5.68	Shelby Township	5.27
Whitmore Lake	5.63	Pontiac	5.20
Armada	5.60	Deckerville	5.00

Freeport, Texas, and moved northward through Texas. It then accelerated toward Michigan, traveling 1,000 miles to near Battle Creek in just twenty-four hours, where it interacted with an approaching cold front. Most hurricanes "spin down" after any appreciable time over land. This storm, however, generated hurricane-force wind gusts (75 mph) over southeast Michigan, with considerable damage reported across the area. The Detroit storm of September 25, 1941, was no hurricane—but it once was!

More recently, Hurricane Ike struck Galveston, Texas, on September 13, 2008, and took a nearly identical path to southern Michigan as did the 1941 storm. Although Ike did not have nearly as much "punch," Detroit did record wind gusts of up to thirty-two miles per hour. More significantly, Ike's rain, combined with the rain from a former Pacific tropical storm named Lowell that struck southern Michigan the previous day, dumped four to six inches of total rain and generated considerable flooding over a large area.

How Do Hurricanes Get Their Names?

It all started back in World War II, when military meteorologists working in the Pacific started naming typhoons (another name for hurricanes) after their girlfriends and wives. This convention stuck, and for many years, lists of women's names were used for hurricanes. Then, lists of alternating men's and women's names were developed in the 1970s as a result of protests during the women's liberation movement.

Currently, there is a six-year rotating list of names, and a name is formally "retired" if that storm becomes particularly catastrophic. So, there will never be another Hurricane Andrew or Hurricane Katrina, at least by name.

So what happens if we get more storms than we have names? This actually happened in 2005 (the year Hurricane Katrina hit). After reaching the last name on the list, the National Hurricane Center started using the Greek alphabet, and we ended the season with tropical storms and hurricanes named Alpha, Beta, Gamma, Delta, Epsilon, and Zeta!

What Causes Rainbows?

Have you ever seen prisms—those triangular-shaped pieces of glass that turn light into different colors? The same effect causes rainbows. Raindrops in the air act as little prisms, bending incoming sunlight and breaking it down into its component colors: red, orange, yellow, green, blue, indigo, and violet.

If you ever want to create a rainbow yourself, take a garden hose on a sunny day and set the nozzle to a fine spray. After a little experimenting, you should be able to orient the spray and yourself to the sun in such a way that you see a spectacular little rainbow (see fig. 16)!

What Causes Fog?

Fog is basically a cloud near the surface and results from any process that saturates the air with moisture, either by cooling the air temperature to the dewpoint or by raising the dewpoint to the air temperature (see fig. 18). This causes very small water droplets to condense out of the air. There are many different types of fog:

1. advection fog—caused by warm air flowing over a colder surface or snow cover
2. freezing fog—forms in below-freezing temperatures
3. ice fog—forms in temperatures much below freezing, where water vapor condenses (subliminates) directly into ice crystals

4. radiation fog—forms on nights with high humidity and good radiational cooling (see the discussion of radiational cooling in chapter 6)
5. rain-induced fog—caused by warm rain falling into a cooler air mass at the ground
6. steam fog—caused by cold air flowing over warmer water
7. upslope fog—caused by air traveling up higher terrain and cooling

What Is Black Ice?

Black ice is a very thin layer of ice that is difficult to see upon casual observation. Black ice here in Michigan typically forms from freezing fog as high moisture content in the air deposits a thin film of moisture on a subfreezing surface, from light rain or melting snow wetting a surface and later freezing, or from freezing drizzle or a brief period of light freezing rain.

Black ice is particularly dangerous due to the sudden, unexpected change in surface traction encountered by pedestrians and motorists. Drivers need to be especially cautious on bridges and overpasses in subfreezing weather when roads appear wet, because these often ice up first due to the road being exposed to cold air temperatures from both above and below.

Do Volcanoes Affect Michigan?

Volcanoes do not directly affect Michigan, but the most explosive eruptions sometimes affect our weather. If a volcano erupts with enough violence, enormous amounts of ash are blown straight up into the stratosphere. These particles stay up there for a year or two and slightly diminish solar radiation. Following Mount Pinatubo's September 1991 eruption in the Philippines, ash traveled around the Northern Hemisphere, creating gorgeous sunsets and something else: the "Bummer Summer of 1992." See table 8 for a sampling of some of the bone-chilling high temperatures from that summer, and many of these set all-time records for the lowest high temperature on these days. Remember, these are daily *high* temperatures!

TABLE 8. Sample of Daily High Temperatures, Summer 1992

JUNE 20, 1992		JULY 30, 1992		AUGUST 28, 1992	
Alpena	46	Milford	63	Houghton Lake	54
Herman	50	Bloomingdale	64	Gladwin	55
Detroit	52	Harbor Beach	64	Lansing	59
Escanaba	53	Lansing	64	Baldwin	61
Houghton Lake	53	Manistique	64	Owosso	61
Trout Lake	53	Detroit	65	Detroit	62
Midland	55	Three Rivers	66	Fayette	62
Boyne Falls	58	Copper Harbor	67	Marquette	62
Three Rivers	58	Gladwin	67	East Tawas	64

What Causes Lake Breezes?

If you live near one of the Great Lakes, you know that a nice cool breeze blows in off the lake on many summer afternoons. This is due to the difference in temperature between land and water, as air heats up much more over land than it does over the water. Once air over land heats up, it starts to rise. Cooler air over the lake then flows in toward land to replace that rising air, and this establishes a little cycle that continues until the land cools enough that the air stops rising. Incidentally, if the land gets significantly cooler than the water, then the reverse cycle happens: air over the water rises, and a land breeze develops that flows out into the lake.

Do lake breezes have any significance other than providing nice, cool relief on a hot summer day? You bet they do. As the cooler air flows in off the lake, the front edge of that air sometimes pushes many miles inland. In essence, the lake breeze generates a mini–cold front. If the atmosphere is unstable enough, a strong lake breeze is all it takes to initiate thunderstorms—even severe storms. That is why we meteorologists carefully watch the progress of lake breezes using surface observations, radar, and high-resolution satellite images on potential thunderstorm days.

What Are Comets?

Comets are relatively small bodies traveling through the solar system. Generally described either as "dirty snowballs" or "snowy dirtballs," a comet's nucleus is comprised of a combination of rocks, dust, ice, and frozen gases. As comets get closer to the sun, melting commences, releasing dust and gases and causing a ball of this material to surround the nucleus. This is called the coma. As the comet gets closer to the sun, the solar wind (charged particles streaming away from the sun) pushes some of the coma's materials millions of miles away from the comet, which we see as the tail. Most people do not realize that comets have two tails: a dust tail and a gas (also called ion) tail. The gas tail is very difficult to see without a telescope, and most people only see the dust tail (see fig. 15).

Some comets travel once through the solar system and never return, while others travel in long elliptical orbits that bring them back at regular intervals. The most famous of these comets is Halley's comet, which returns every seventy-six years and will make its next appearance in 2061.

I consider comets to be the most fascinating and important of all objects to study in space, as they are pristine, leftover matter from the origin of the solar system. It is thought that a comet bombardment nearly four billion years ago brought most of our water to Earth, and some scientists believe that comets even brought the building blocks for life. Thus, by studying comets, we learn more about Earth's origin and, possibly, about the origin of life itself on Earth.

What Are "Shooting Stars"?

Shooting stars, more accurately called meteors, are small bits of rocky material from space that hit Earth's atmosphere at a speed of between twenty-five and forty-five miles per second. Once they get within about seventy miles of Earth, meteors encounter increasing atmospheric friction as they streak through denser and denser air, and they heat up to the point of burning up in a bright streak of light. Have you ever had Grape Nuts cereal? Most meteors are only the size of a single "grape nut" and completely burn up. Here are peak dates for the most popular annual meteor showers, which result from Earth passing through debris trails left by passing comets many years ago:

Quadrantids	January 4
Perseids	August 12
Orionids	October 21
Leonids	November 17
Geminids	December 14

Occasionally, larger space rocks that do not completely burn up eventually reach Earth's surface, and when they do, they are called meteorites. If you ever take a trip to Arizona, I strongly suggest visiting Meteor Crater near Flagstaff. It's a 4,000-foot-diameter crater created by a 150-foot meteorite that hit 50,000 years ago and is an amazing site to see.

What Are the Northern Lights?

The Northern Hemisphere's northern lights, more properly called the aurora borealis (compare the aurora australis in the Southern Hemisphere), begin with enormous eruptions on the sun. These *coronal mass ejections* (CMEs) accelerate the normal stream of charged particles flowing out of the sun toward Earth, and these particles interact with gas atoms high in our atmosphere. When this happens, the gas atoms become "excited" and shed their extra energy in the form of light.

Auroras are more common in northern Michigan than in the south, due to that fact that these atomic light shows are more prominent at higher latitudes. However, if a powerful CME occurs near the sun's equator when that part of the sun directly faces Earth, the charged particles are "aimed" right at us, thus forcing the auroral ring around our planet further south. The more powerful the CME, the further south the aurora travels.

No two auroras are alike: some look like delicate curtains of color across the sky, some take on a flowing mysterious look, and others feature multiple colors. But there is no mistaking the beautiful pattern of color when the northern lights put on a show (see fig. 19).

How Did the National Weather Service (NWS) Get Started?

The NWS traces its formal beginnings to a joint congressional resolution signed into law by President Ulysses S. Grant in 1870, which ordered the organization of a group of trained weather observers across the country "to provide for taking meteorological observations at the military stations in the interior of the continent and at other points in the States and Territories . . . and for giving notice on the northern [Great] Lakes and on the seacoast by magnetic telegraph and marine signals, of the approach and force of storms." The secretary of war oversaw development of the Weather Bureau, because it was decided that military discipline would be needed to ensure consistency and accuracy of the observations.

In 1890, the Weather Bureau became a civilian agency when it was transferred from the Department of War to the Department of Agriculture, and there is a significant Michigan connection: one year later, Professor Mark W. Harrington, director of the University of Michigan's Detroit Observatory, replaced Major General Adolphus Greely as the first civilian head of the Weather Bureau. In 1940, the Weather Bureau was transferred to its current home in the Department of Commerce and was renamed the National Weather Service in 1967.

Do Television Meteorologists Really Do the Weather in Front of a Green Screen?

Yes we do! The special effect many people ask about is called *chroma-key*. What happens is that the studio camera is specially programmed so that it cannot see that particular shade of green. So, the camera can only "see" the weathercaster standing there. Then, the control room electronically replaces what the camera cannot see (the green screen) with whatever video source we want: weather computer, radar, videotape, and so on. In the studio, all we see is the green screen behind us, but you at home see a weather map (see fig. 17).

So how do we know what to point at? That's the "acting" part of presenting the weather on television. Just off to our left and right, but out of camera range, are two television monitors showing what you see at home (there is also a television monitor on the front of the studio camera). So, when we turn our backs to you and face the map, it appears that we are looking at what we are pointing at. In reality, our eyes are directed at the television monitor off to the side, and that shows us where to point. Now you know the secret of television weather presentations!

9 Michigan Weather Extremes

Extreme weather happens everywhere, not just at the primary airport where your local television meteorologist cites the statistics. Fortunately, the National Weather Service has a network of cooperative weather observers throughout the state—ordinary citizens who record weather observations every day and report that data to the NWS. In preparing this chapter, I examined the period of record for all of the state's cooperative weather observers and first identified the stations with the longest-running climate records. From this list, I then selected enough stations to represent as many parts of Michigan as possible. Since the data was not available in an easy-to-use spreadsheet database, I had to manually transcribe each location's daily record highs, lows, precipitation, and snowfall into the charts that follow, which entailed hand typing over 30,000 individual statistics. This was a most time-consuming process, but it resulted in a much closer examination of the data and, therefore, the identification of many errors that I worked with the NWS and NCDC to correct. There is no question that the end justified the means. Here are some things you need to know before using the statistics.

1. A particular station's length of climate record is not necessarily the number of years used for a particular extreme, due to missing data. For example, Traverse City has a 114-year climate record, but its temperature extremes are based on 110 years of records, its precipitation on 110 years of records, and its snowfall on 101 years of records.
2. Record-breaking weather is defined as the most extreme *nonmissing* value in a climate record. So, if official data is missing from a day on which extreme weather occurred, then the next most extreme data for that date is considered the record. For example, Munising has a record snowfall of seven inches on February 29. However, Munising's record precipitation for that day is only 0.10 inch. How can this be? Although the official snowfall measurement exists in the database, the precipitation measurement for that same day is missing.
3. Keep in mind that February 29 only comes around once every four years, so records for this date will in many instances be quite different than on surrounding days.
4. You also may notice significant differences in weather extremes for the same day at relatively close reporting stations, and this is most likely due to differences in the length of the climate record. For example, Detroit's December 22 record low temperature is –24 degrees, but Pontiac's record low for that date is only –6 degrees (and Pontiac is usually colder than Detroit). The reason for this discrepancy is that Detroit's record was set in 1872, well before the Pontiac database starts.
5. Only *measurable* snowfall is considered a candidate for a daily extreme. A trace of snow (less than 0.1 inch) is not represented in the charts. Thus, the "first" snow of the season is the first measurable snowfall in the following charts. There certainly may have been previous dates with snowflakes in the air, but only measurable snow is considered for the daily records.
6. Each day's weather extreme is for a period of twenty-four hours. So, an eighteen-inch snowfall that you remember for a particular date but that is not reflected in the chart may have fallen over two

different observing dates. Also keep in mind that a different amount of snow may have fallen at the cooperative weather observer's location than at your location.

7. The earliest 32-degree temperature on the chart is not necessarily the first frost where you live. Changes in terrain and elevation impact temperatures, and a particular reporting station's earliest frost may be earlier or later than the first frost in your location.
8. The year's most extreme value in each category is highlighted in bold type.
9. The weather extremes that follow are valid through December 2009. Although I made every effort to identify errors and notify the NWS and NCDC to have them checked and corrected, it is unrealistic to expect that I found every error. If you feel that any of these statistics are in error, please contact me at ExtremeMichiganWeather@GrossWeather.com, and I will check them and publish corrections in a future edition of this book.
10. If you feel you have the dedication to become an NWS cooperative weather observer, then contact the NWS office closest to you through its Web site:

Detroit/Pontiac	www.crh.noaa.gov/dtx
Northern Indiana	www.crh.noaa.gov/iwx
Grand Rapids	www.crh.noaa.gov/grr
Gaylord	www.crh.noaa.gov/apx
Marquette	www.crh.noaa.gov/mqt

11. Another opportunity to help monitor the nation's precipitation climatology is by assisting the Community Collaborative Rain, Hail and Snow Network (CoCoRaHS). This grassroots effort hopes to have as many as 25,000 volunteers measuring rain, snow, and hail by the end of 2010, with the ultimate goal of one or more observers per square mile in heavily populated urban and suburban areas and at least one observer in each rural town. For more information about CoCoRaHS, check out its Web site at www.cocorahs.org.

Climate Records

Adrian | 76

Alpena | 79

Ann Arbor | 83

Benton Harbor | 87

Big Rapids | 91

Bloomingdale | 94

Coldwater | 98

Detroit | 102

East Tawas | 106

Flint | 109

Gaylord | 113

Gladwin | 117

Grand Rapids | 121

Hancock | 124

Harbor Beach | 128

Ironwood | 132

Jackson | 136

Munising | 139

Pontiac | 143

Port Huron | 147

Sault Ste. Marie | 151

Traverse City | 154

Adrian (Climate Record Begins 1887)

	RECORD HIGH	RECORD LOW	RECORD PRECIPITATION	RECORD SNOWFALL
January 1	54	–8	1.64″	5.0″
January 2	57	–8	1.30″	10.0″
January 3	59	–15	0.85″	3.5″
January 4	62	–21	1.93″	4.0″
January 5	59	–15	1.41″	4.2″
January 6	60	–5	0.60″	6.0″
January 7	62	–8	0.88″	2.5″
January 8	62	–18	1.62″	4.0″
January 9	57	–11	0.70″	5.5″
January 10	58	–13	0.78″	7.0″
January 11	66	–17	0.63″	6.0″
January 12	62	–24	2.00″	5.0″
January 13	62	–17	0.81″	7.5″
January 14	56	–12	0.92″	7.0″
January 15	62	–18	0.72″	2.5″
January 16	53	–17	0.60″	2.5″
January 17	64	–16	1.20″	2.5″
January 18	63	–18	1.16″	7.5″
January 19	56	–22	0.70″	5.5″
January 20	66	**–26**	1.15″	4.0″
January 21	65	–21	1.07″	3.5″
January 22	64	–11	1.34″	6.5″
January 23	61	–17	0.65″	4.0″
January 24	62	–17	1.21″	5.5″
January 25	68	–23	0.60″	6.0″
January 26	66	–16	1.46″	**15.0″**
January 27	64	–10	0.60″	4.5″
January 28	58	–12	0.79″	3.7″
January 29	59	–15	1.50″	4.0″
January 30	52	–11	1.22″	9.3″
January 31	60	–12	2.25″	9.8″
February 1	56	–11	1.18″	5.0″
February 2	57	–10	0.61″	3.0″
February 3	53	–16	0.87″	8.8″
February 4	64	–13	1.38″	4.5″
February 5	52	–24	1.19″	6.0″
February 6	58	–8	0.94″	6.0″
February 7	54	–10	1.00″	9.0″
February 8	64	–13	1.18″	6.0″
February 9	58	–16	0.97″	4.0″
February 10	56	–21	1.05″	7.3″
February 11	70	–10	1.03″	5.0″
February 12	61	–19	1.00″	5.0″
February 13	65	–17	1.37″	5.5″
February 14	64	–17	1.40″	5.0″
February 15	66	–15	0.63″	3.5″
February 16	62	–9	1.46″	4.2″
February 17	59	–18	1.89″	3.5″
February 18	57	–10	0.63″	4.0″
February 19	62	–7	1.86″	3.0″
February 20	64	–21	0.95″	8.0″
February 21	65	–8	1.17″	8.0″
February 22	65	–8	2.35″	9.5″
February 23	64	–11	1.46″	4.0″
February 24	64	–11	0.80″	8.0″
February 25	68	–10	1.00″	8.5″
February 26	69	–11	1.20″	12.0″
February 27	64	–12	0.82″	5.5″
February 28	60	–10	1.50″	**15.0″**
February 29	60	2	2.00″	2.1″
March 1	64	0	1.31″	3.5″
March 2	62	–6	0.50″	3.5″
March 3	68	–6	1.95″	8.0″
March 4	72	2	1.14″	8.0″
March 5	70	–7	0.94″	7.0″
March 6	70	–2	0.80″	4.5″
March 7	75	–6	1.20″	2.0″
March 8	79	–8	1.53″	6.0″
March 9	67	–4	1.07″	6.5″
March 10	72	2	1.52″	5.0″
March 11	71	3	0.85″	2.5″
March 12	76	–2	1.37″	3.0″
March 13	73	2	1.60″	2.5″
March 14	78	1	1.51″	6.0″
March 15	77	0	0.90″	9.0″
March 16	76	–3	0.95″	3.0″
March 17	75	–3	1.13″	11.5″
March 18	80	3	1.08″	2.8″
March 19	80	1	1.89″	3.0″
March 20	76	8	0.84″	6.0″
March 21	76	6	1.10″	8.0″
March 22	82	–2	1.55″	11.5″
March 23	80	–7	1.26″	4.0″
March 24	84	3	1.12″	1.0″
March 25	83	2	1.55″	3.0″
March 26	80	7	1.28″	4.0″
March 27	85	6	1.19″	9.3″
March 28	85	–8	1.61″	3.5″
March 29	83	11	1.30″	12.0″
March 30	80	12	1.58″	3.0″
March 31	80	6	1.57″	2.0″
April 1	77	10	1.30″	2.5″
April 2	82	18	1.08″	1.5″
April 3	78	11	1.40″	1.0″
April 4	78	13	1.20″	0.5″
April 5	81	10	2.12″	8.0″
April 6	86	14	0.75″	6.5″
April 7	85	8	0.90″	7.0″
April 8	79	9	1.03″	2.0″

Adrian *(continued)*

	RECORD HIGH	RECORD LOW	RECORD PRECIPITATION	RECORD SNOWFALL
April 9	79	16	1.03″	3.5″
April 10	83	16	1.36″	1.0″
April 11	88	20	1.84″	3.0″
April 12	88	15	1.06″	0.6″
April 13	85	18	1.17″	0.5″
April 14	83	19	1.17″	1.0″
April 15	84	19	1.48″	1.5″
April 16	88	21	0.99″	3.0″
April 17	88	21	1.18″	3.0″
April 18	87	18	1.33″	0.5″
April 19	85	23	2.05″	0.5″
April 20	86	17	2.13″	3.0″
April 21	86	20	1.27″	4.0″
April 22	87	22	1.03″	
April 23	89	17	1.37″	0.5″
April 24	90	24	1.42″	
April 25	90	24	1.47″	
April 26	92	23	1.38″	
April 27	86	25	1.33″	
April 28	86	26	1.98″	
April 29	88	26	2.50″	
April 30	89	28	1.06″	
May 1	90	20	1.82″	
May 2	90	22	1.55″	
May 3	90	25	1.95:	
May 4	91	25	1.00″	0.5″
May 5	95	26	1.91″	
May 6	91	29	1.66″	
May 7	89	26	0.70″	
May 8	91	28	1.52″	
May 9	92	29	2.34″	4.0″
May 10	92	23	1.80″	0.5″
May 11	92	28	2.50″	
May 12	88	29	2.81″	
May 13	89	29	1.48″	
May 14	89	28	2.00″	
May 15	92	28	1.37″	
May 16	92	30	2.02″	
May 17	94	27	1.95″	
May 18	95	29	3.00″	
May 19	95	30	2.20″	
May 20	93	29	0.88″	
May 21	93	26	1.14″	
May 22	91	30	1.14″	
May 23	90	31	1.41″	
May 24	91	33	1.72″	
May 25	92	31	1.72″	
May 26	98	35	3.16″	
May 27	99	32	2.01″	
May 28	98	29	1.57″	
May 29	97	34	3.57″	
May 30	96	34	1.66″	
May 31	97	34	1.72″	
June 1	103	35	1.20″	
June 2	100	36	2.32″	
June 3	100	34	1.68″	
June 4	98	36	1.68″	
June 5	99	38	1.15″	
June 6	98	37	3.60″	
June 7	99	40	1.40″	
June 8	100	37	1.75″	
June 9	98	34	1.05″	
June 10	98	33	1.15″	
June 11	100	34	1.00″	
June 12	96	39	2.00″	
June 13	95	41	1.96″	
June 14	97	37	1.69″	
June 15	98	41	0.98″	
June 16	103	40	3.25″	
June 17	98	40	2.25″	
June 18	99	42	1.86″	
June 19	98	42	1.68″	
June 20	100	38	2.27″	
June 21	101	37	3.20″	
June 22	96	37	1.65″	
June 23	99	39	4.57″	
June 24	99	41	2.61″	
June 25	104	37	3.12″	
June 26	100	40	1.30″	
June 27	102	43	1.73″	
June 28	106	43	2.88″	
June 29	103	45	1.90″	
June 30	102	41	1.76″	
July 1	101	41	1.40″	
July 2	101	41	3.10″	
July 3	102	44	1.53″	
July 4	101	42	2.73″	
July 5	102	43	0.95″	
July 6	99	43	1.37″	
July 7	100	43	2.90″	
July 8	105	44	2.82″	
July 9	106	44	2.05″	
July 10	105	44	1.22″	
July 11	103	41	1.90″	
July 12	106	47	1.64″	
July 13	106	43	1.22″	
July 14	**108**	44	1.54″	
July 15	100	43	1.90″	
July 16	100	45	1.70″	
July 17	103	45	2.30″	
July 18	101	47	1.30″	

Adrian *(continued)*

	RECORD HIGH	RECORD LOW	RECORD PRECIPITATION	RECORD SNOWFALL
July 19	105	45	2.58″	
July 20	105	44	1.10″	
July 21	102	44	2.00″	
July 22	103	46	3.61″	
July 23	103	44	1.44″	
July 24	**108**	44	1.35″	
July 25	105	45	3.35″	
July 26	100	43	1.79″	
July 27	103	44	2.05″	
July 28	101	49	1.50″	
July 29	103	47	1.25″	
July 30	100	46	1.78″	
July 31	102	42	1.99″	
August 1	100	44	1.95″	
August 2	100	45	1.46″	
August 3	102	40	1.61″	
August 4	101	39	2.20″	
August 5	104	42	1.72″	
August 6	107	41	1.80″	
August 7	103	44	2.72″	
August 8	103	43	1.20″	
August 9	102	42	2.02″	
August 10	100	43	1.15″	
August 11	99	45	1.77″	
August 12	100	39	2.79″	
August 13	101	39	3.30″	
August 14	100	43	2.93″	
August 15	98	42	1.42″	
August 16	99	40	2.22″	
August 17	98	44	2.00″	
August 18	100	42	1.90″	
August 19	101	41	2.70″	
August 20	102	40	2.00″	
August 21	104	40	2.09″	
August 22	106	41	1.06″	
August 23	100	38	2.08″	
August 24	100	43	1.68″	
August 25	99	42	0.93″	
August 26	100	44	4.20″	
August 27	101	42	2.30″	
August 28	95	38	2.00″	
August 29	95	32	0.98″	
August 30	99	32	1.10″	
August 31	100	36	1.29″	
September 1	98	39	1.50″	
September 2	104	35	0.80″	
September 3	100	36	**4.74**″	
September 4	94	38	3.80″	
September 5	99	40	2.05″	
September 6	95	36	2.10″	
September 7	99	31	1.57″	
September 8	100	29	1.18″	
September 9	96	30	1.63″	
September 10	99	37	1.79″	
September 11	100	35	1.45″	
September 12	98	37	2.75″	
September 13	96	33	2.83″	
September 14	98	29	1.37″	
September 15	100	31	1.14″	
September 16	98	35	1.94″	
September 17	95	32	2.70″	
September 18	93	35	1.98″	
September 19	96	31	1.86″	
September 20	96	30	1.53″	
September 21	97	28	2.63″	
September 22	97	28	1.65″	
September 23	95	27	1.56″	
September 24	95	30	0.92″	
September 25	95	32	1.85″	
September 26	92	27	1.25″	
September 27	92	29	1.45″	
September 28	94	27	1.50″	
September 29	95	26	2.00″	
September 30	89	29	1.50″	
October 1	90	21	2.08″	
October 2	89	26	1.26″	
October 3	90	24	1.83″	
October 4	91	25	0.93″	
October 5	89	26	1.46″	
October 6	91	24	2.82″	
October 7	89	24	1.73″	
October 8	91	18	1.31″	
October 9	87	25	1.40″	
October 10	88	22	1.13″	
October 11	85	20	2.31″	
October 12	85	22	2.63″	
October 13	82	25	1.32″	
October 14	86	22	1.57″	
October 15	89	23	2.34″	
October 16	84	24	1.79″	
October 17	84	21	1.88″	
October 18	85	20	1.58″	
October 19	85	22	1.34″	3.0″
October 20	84	19	1.12″	1.0″
October 21	83	17	0.91″	
October 22	84	23	1.17″	
October 23	83	18	1.97″	
October 24	81	19	1.26″	
October 25	81	20	1.14″	
October 26	82	20	0.88″	
October 27	76	19	1.12″	

Adrian *(continued)*

	RECORD HIGH	RECORD LOW	RECORD PRECIPITATION	RECORD SNOWFALL
October 28	79	21	0.66″	1.0″
October 29	76	19	0.90″	
October 30	77	15	1.11″	
October 31	80	15	0.92″	
November 1	80	15	2.02″	
November 2	77	14	1.84″	4.0″
November 3	76	16	1.24″	3.0″
November 4	76	9	1.20″	1.0″
November 5	74	6	1.50″	1.5″
November 6	74	13	1.58″	8.0″
November 7	73	8	1.83″	3.0″
November 8	72	11	1.13″	0.5″
November 9	74	12	1.37″	3.0″
November 10	67	16	1.11″	0.9″
November 11	71	17	1.02″	3.0″
November 12	71	14	1.70″	4.5″
November 13	71	7	1.03″	5.5″
November 14	72	12	0.75″	1.5″
November 15	70	11	2.00″	2.0″
November 16	70	6	2.19″	8.0″
November 17	70	9	1.02″	3.7″
November 18	68	9	1.27″	2.5″
November 19	70	10	1.35″	3.0″
November 20	72	9	1.15″	1.5″
November 21	70	10	1.18″	1.8″
November 22	70	10	1.70″	2.0″
November 23	71	5	1.13″	3.5″
November 24	67	0	0.91″	3.5″
November 25	64	3	1.80″	4.5″
November 26	66	6	1.68″	5.5″
November 27	68	6	2.32″	5.0″
November 28	67	4	1.25″	3.8″
November 29	65	3	1.30″	4.5″
November 30	65	–5	2.60″	2.6″

	RECORD HIGH	RECORD LOW	RECORD PRECIPITATION	RECORD SNOWFALL
December 1	65	5	1.08″	14.0″
December 2	68	–8	1.00″	10.5″
December 3	67	–8	1.08″	3.0″
December 4	64	–2	0.93″	4.0″
December 5	69	–1	1.60″	5.5″
December 6	67	–3	0.73″	6.0″
December 7	65	1	1.63″	5.0″
December 8	64	–5	1.30″	7.0″
December 9	60	–6	0.85″	4.3″
December 10	66	–4	1.35″	7.0″
December 11	62	–11	1.16″	5.5″
December 12	61	–11	1.70″	3.0″
December 13	60	–12	1.84″	4.5″
December 14	66	–13	1.71″	4.5″
December 15	62	–17	1.52″	6.0″
December 16	64	–11	0.73″	4.5″
December 17	57	–9	1.60″	3.8″
December 18	56	–6	0.88″	5.0″
December 19	57	–6	1.10″	10.0″
December 20	59	–5	1.00″	5.0″
December 21	61	–10	1.93″	3.5″
December 22	57	–13	0.84″	4.0″
December 23	58	–12	0.87″	7.0″
December 24	63	–16	2.17″	3.0″
December 25	63	–12	1.02″	8.5″
December 26	63	–14	0.70″	6.0″
December 27	60	–15	1.25″	4.0″
December 28	62	–12	0.99″	4.0″
December 29	64	–6	1.79″	4.0″
December 30	60	–10	1.55″	5.0″
December 31	60	–10	1.28″	5.5″

Alpena (Climate Record Begins 1887)

	RECORD HIGH	RECORD LOW	RECORD PRECIPITATION	RECORD SNOWFALL
January 1	56	–18	0.77″	9.0″
January 2	50	–9	0.71″	5.6″
January 3	52	–21	0.80″	8.4″
January 4	48	–19	1.23″	7.4″
January 5	50	–16	0.54″	3.8″
January 6	54	–16	0.77″	8.5″
January 7	48	–13	1.41″	8.2″
January 8	50	–21	0.57″	3.0″
January 9	46	–19	0.49″	7.7″
January 10	52	–21	0.63″	5.8″
January 11	48	–20	0.38″	5.1″

	RECORD HIGH	RECORD LOW	RECORD PRECIPITATION	RECORD SNOWFALL
January 12	49	–16	0.50″	10.0″
January 13	51	–17	0.79″	8.5″
January 14	47	–17	1.24″	3.8″
January 15	48	–28	0.44″	5.7″
January 16	51	–28	0.33″	3.9″
January 17	50	–19	0.52″	5.1″
January 18	52	–23	0.57″	7.4″
January 19	51	–20	0.97″	4.8″
January 20	45	–18	0.81″	9.0″
January 21	45	–21	0.50″	10.0″
January 22	46	–16	0.86″	7.8″

Alpena *(continued)*

	RECORD HIGH	RECORD LOW	RECORD PRECIPITATION	RECORD SNOWFALL
January 23	49	–23	0.41″	4.6″
January 24	44	–17	0.81″	4.5″
January 25	52	–19	0.78″	4.7″
January 26	62	–22	1.83″	16.3″
January 27	55	–26	1.11″	11.2″
January 28	48	–22	0.80″	8.0″
January 29	49	–18	0.88″	4.7″
January 30	47	–26	1.00″	10.0″
January 31	50	–25	0.67″	6.1″
February 1	49	–27	0.69″	7.5″
February 2	46	–27	0.50″	4.2″
February 3	57	–22	0.58″	4.9″
February 4	53	–24	0.64″	7.8″
February 5	48	–27	0.69″	12.8″
February 6	46	–16	0.64″	4.6″
February 7	51	–23	0.54″	6.5″
February 8	50	–34	0.55″	6.8″
February 9	52	–36	0.45″	3.6″
February 10	55	–25	0.71″	11.2″
February 11	59	–24	0.91″	6.5″
February 12	58	–21	0.61″	6.4″
February 13	49	–19	0.69″	7.0″
February 14	52	–17	0.81″	7.3″
February 15	52	–27	0.83″	8.4″
February 16	56	–25	0.37″	6.7″
February 17	57	**–37**	0.55″	4.2″
February 18	54	–34	0.49″	7.1″
February 19	58	–20	0.81″	6.5″
February 20	54	–31	1.21″	4.9″
February 21	50	–18	1.51″	6.1″
February 22	60	–24	2.86″	**18.2**″
February 23	65	–13	0.73″	11.4″
February 24	56	–24	0.68″	7.8″
February 25	58	–24	0.94″	8.3″
February 26	59	–18	0.55″	6.4″
February 27	54	–14	1.09″	3.5″
February 28	52	–13	0.89″	8.0″
February 29	55	–12	0.40″	4.0″
March 1	48	–23	1.19″	6.2″
March 2	58	–27	0.76″	5.5″
March 3	63	–21	0.48″	12.2″
March 4	55	–27	0.93″	17.3″
March 5	49	–22	0.97″	4.1″
March 6	61	–17	0.91″	9.1″
March 7	73	–12	1.19″	12.1″
March 8	80	–20	0.85″	8.5″
March 9	67	–17	1.03″	11.6″
March 10	61	–9	0.91″	7.0″
March 11	62	–15	1.52″	3.6″
March 12	66	–13	0.82″	8.3″
March 13	71	–10	0.87″	7.1″
March 14	75	–5	1.06″	6.3
March 15	76	–6	0.76″	4.2″
March 16	69	–10	0.90″	3.2″
March 17	74	–8	1.46″	12.2″
March 18	65	–12	0.77″	3.4″
March 19	63	–15	1.14″	9.7″
March 20	69	–12	0.87″	3.7″
March 21	68	–5	0.74″	10.8″
March 22	79	–9	0.60″	4.3″
March 23	70	–15	0.78″	4.3″
March 24	66	–4	0.95″	10.0″
March 25	69	–14	1.15″	14.0″
March 26	76	–9	0.89″	10.0″
March 27	75	3	0.62″	4.0″
March 28	82	–7	0.50″	3.0″
March 29	75	–8	0.77″	3.2″
March 30	74	2	0.85″	5.3″
March 31	74	–10	1.85″	14.2″
April 1	74	–6	1.19″	12.5″
April 2	67	0	0.93″	7.5″
April 3	81	3	0.94″	5.0″
April 4	79	0	0.75″	5.6″
April 5	81	5	0.88″	5.5″
April 6	81	–7	0.86″	3.0″
April 7	86	5	0.87″	3.9″
April 8	74	7	1.17″	3.5″
April 9	77	4	1.34″	2.8″
April 10	81	7	1.10″	10.8″
April 11	83	13	1.15″	1.6″
April 12	81	10	1.27″	5.3″
April 13	81	9	1.49″	3.4″
April 14	81	14	1.05″	3.4″
April 15	82	11	0.88″	8.0″
April 16	90	18	0.85″	5.3″
April 17	85	10	0.74″	3.9″
April 18	85	16	1.00″	4.0″
April 19	83	9	1.64″	3.5″
April 20	86	18	0.90″	2.8″
April 21	89	19	0.88″	2.0″
April 22	88	17	1.12″	1.5″
April 23	87	18	0.95″	5.0″
April 24	85	20	1.32″	3.5″
April 25	90	20	1.27″	0.4″
April 26	89	18	0.75″	1.6″
April 27	88	19	1.22″	2.8″
April 28	85	20	1.30″	
April 29	80	20	0.50″	1.9″
April 30	87	22	1.11″	4.3″
May 1	84	23	1.20″	0.9″

Alpena *(continued)*

	RECORD HIGH	RECORD LOW	RECORD PRECIPITATION	RECORD SNOWFALL
May 2	86	20	1.25″	0.7″
May 3	86	20	1.16″	8.7″
May 4	88	21	1.36″	0.6″
May 5	92	22	0.57″	1.8″
May 6	89	23	2.21″	3.7″
May 7	92	20	2.11″	1.0″
May 8	89	22	1.87″	0.2″
May 9	89	22	2.01″	1.6″
May 10	88	22	1.18″	0.3″
May 11	86	21	0.86″	
May 12	87	23	1.13″	
May 13	86	22	0.74″	
May 14	87	24	1.73″	
May 15	92	23	1.01″	
May 16	94	26	1.14″	
May 17	91	23	1.35″	
May 18	90	22	0.71″	
May 19	88	25	1.03″	
May 20	92	24	1.09″	
May 21	91	27	1.04″	
May 22	91	26	1.58″	4.0″
May 23	89	26	1.46″	
May 24	88	24	0.80″	
May 25	87	26	1.57″	
May 26	90	25	1.34″	
May 27	93	27	1.98″	
May 28	93	27	1.61″	0.3″
May 29	92	29	1.70″	
May 30	94	29	0.88″	
May 31	92	29	2.00″	
June 1	104	28	0.81″	
June 2	90	27	1.05″	
June 3	92	27	1.29″	
June 4	93	27	1.05″	
June 5	95	29	0.86″	
June 6	96	31	1.12″	
June 7	92	31	1.32″	
June 8	96	29	1.29″	
June 9	93	29	0.61″	
June 10	93	31	1.55″	
June 11	93	30	1.08″	
June 12	97	33	2.52″	
June 13	97	31	1.78″	
June 14	98	31	1.84″	
June 15	95	32	1.25″	
June 16	96	31	0.74″	
June 17	95	34	1.51″	
June 18	97	31	1.06″	
June 19	103	34	1.29″	
June 20	98	31	1.12″	
June 21	93	32	0.73″	

	RECORD HIGH	RECORD LOW	RECORD PRECIPITATION	RECORD SNOWFALL
June 22	93	34	1.42″	
June 23	94	35	0.95″	
June 24	97	30	1.12″	
June 25	95	34	1.56″	
June 26	98	35	2.32″	
June 27	99	33	1.05″	
June 28	98	32	1.47″	
June 29	99	35	1.85″	
June 30	99	36	2.21″	
July 1	100	34	1.46″	
July 2	98	37	0.60″	
July 3	97	37	2.80″	
July 4	98	34	2.12″	
July 5	96	37	2.59″	
July 6	100	37	0.99″	
July 7	100	36	1.53″	
July 8	104	38	2.74″	
July 9	102	39	1.69″	
July 10	104	41	1.77″	
July 11	100	37	0.76″	
July 12	105	39	1.04″	
July 13	**106**	37	1.57″	
July 14	102	40	0.82″	
July 15	98	39	1.02″	
July 16	98	40	0.85″	
July 17	95	38	2.14″	
July 18	96	40	1.19″	
July 19	93	39	2.01″	
July 20	98	34	1.82″	
July 21	96	41	1.68″	
July 22	97	42	1.73″	
July 23	95	41	1.47″	
July 24	96	40	1.43″	
July 25	95	40	1.27″	
July 26	98	39	2.34″	
July 27	96	39	1.16″	
July 28	97	44	0.66″	
July 29	99	36	1.64″	
July 30	98	36	2.38″	
July 31	100	35	1.10″	
August 1	98	38	2.61″	
August 2	102	35	1.44″	
August 3	98	39	0.86″	
August 4	95	38	1.60″	
August 5	97	37	0.87″	
August 6	100	34	1.19″	
August 7	97	40	1.73″	
August 8	97	40	0.83″	
August 9	96	38	1.31″	
August 10	94	39	2.01″	

Alpena *(continued)*

	RECORD HIGH	RECORD LOW	RECORD PRECIPITATION	RECORD SNOWFALL
August 11	92	37	1.63″	
August 12	97	40	1.89″	
August 13	95	35	1.25″	
August 14	95	35	1.38″	
August 15	94	36	1.20″	
August 16	93	35	1.17″	
August 17	92	33	0.86″	
August 18	100	34	1.10″	
August 19	98	35	1.56″	
August 20	95	34	1.74″	
August 21	99	31	1.79″	
August 22	97	32	1.85″	
August 23	95	30	1.69″	
August 24	100	32	1.94″	
August 25	100	36	1.11″	
August 26	94	32	1.15″	
August 27	94	36	1.63″	
August 28	96	31	1.84″	
August 29	95	30	1.71″	
August 30	96	29	1.95″	
August 31	94	31	2.96″	
September 1	99	36	1.17″	
September 2	97	31	2.01″	
September 3	98	31	**5.14**″	
September 4	94	32	1.02″	
September 5	90	31	1.17″	
September 6	94	30	1.02″	
September 7	93	34	0.86″	
September 8	94	33	0.59″	
September 9	95	30	1.64″	
September 10	94	32	3.02″	
September 11	98	28	1.17″	
September 12	97	28	1.80″	
September 13	92	28	1.83″	
September 14	94	26	1.65″	
September 15	95	26	1.17″	
September 16	90	26	1.24″	
September 17	92	28	1.00″	
September 18	93	27	1.35″	
September 19	88	27	1.21″	
September 20	87	24	0.96″	
September 21	88	28	2.25″	
September 22	87	27	1.13″	
September 23	89	23	1.33″	
September 24	89	27	1.95″	0.2″
September 25	85	25	1.44″	
September 26	88	25	1.54″	
September 27	83	25	0.91″	
September 28	84	23	1.06″	0.2″
September 29	84	25	1.20″	
September 30	86	24	2.04″	

	RECORD HIGH	RECORD LOW	RECORD PRECIPITATION	RECORD SNOWFALL
October 1	88	23	0.71″	
October 2	84	24	0.65″	
October 3	84	21	1.30″	0.4″
October 4	86	26	1.59″	1.3″
October 5	85	19	0.54″	
October 6	86	22	3.44″	
October 7	87	21	1.37″	
October 8	90	21	0.87″	
October 9	83	21	0.59″	1.0″
October 10	86	17	1.47″	
October 11	87	21	0.49″	0.8″
October 12	85	21	1.70″	0.2″
October 13	84	19	1.46″	0.5″
October 14	82	20	1.16″	
October 15	84	23	0.95″	
October 16	82	20	1.80″	1.5″
October 17	86	22	0.96″	2.8″
October 18	85	20	1.47″	1.8″
October 19	78	20	1.02″	2.9″
October 20	82	20	0.61″	1.7″
October 21	78	20	0.59″	1.9″
October 22	83	20	2.12″	2.4″
October 23	83	18	1.03″	0.7″
October 24	83	19	1.82″	4.4″
October 25	77	19	0.60″	0.9″
October 26	78	15	0.92″	2.7″
October 27	79	16	1.12″	1.4″
October 28	75	12	0.57″	1.0″
October 29	73	16	0.46″	1.5″
October 30	76	12	1.76″	3.0″
October 31	77	17	0.54″	0.3″
November 1	77	20	1.34″	2.1″
November 2	75	19	1.28″	1.4″
November 3	74	13	1.50″	14.0″
November 4	75	8	1.02″	1.3″
November 5	76	7	1.35″	7.1″
November 6	71	1	0.97″	0.9″
November 7	65	12	2.30″	7.0″
November 8	69	12	0.69″	4.9″
November 9	75	13	1.58″	4.8″
November 10	67	13	1.12″	3.2″
November 11	66	8	0.84″	7.2″
November 12	64	7	0.50″	5.0″
November 13	63	7	0.68″	3.7″
November 14	66	3	0.61″	6.4″
November 15	65	–6	1.06″	3.0″
November 16	69	0	0.81″	9.9″
November 17	72	6	1.13″	2.5″
November 18	69	9	1.13″	3.2″
November 19	72	8	1.15″	3.6″
November 20	65	8	1.50″	3.8″

Alpena *(continued)*

	RECORD HIGH	RECORD LOW	RECORD PRECIPITATION	RECORD SNOWFALL
November 21	65	9	0.74″	4.0″
November 22	60	3	1.09″	5.5″
November 23	65	0	0.65″	6.9″
November 24	62	0	0.55″	4.7″
November 25	59	−1	1.04″	8.0″
November 26	62	−2	0.93″	3.7″
November 27	67	−1	1.78″	6.8″
November 28	60	−2	1.00″	3.8″
November 29	61	−6	1.82″	15.0″
November 30	63	3	0.52″	4.9″
December 1	59	−2	1.28″	6.6″
December 2	60	−6	0.86″	11.9″
December 3	64	−9	1.02″	8.4″
December 4	61	0	0.61″	4.0″
December 5	65	−3	0.91″	8.2″
December 6	62	−3	0.68″	7.4″
December 7	55	−6	1.04″	4.6″
December 8	57	−8	0.78″	7.0″
December 9	59	−4	0.69″	8.0″
December 10	56	−9	1.63″	5.5″
December 11	48	−18	0.64″	3.2″
December 12	53	−11	0.83″	7.5″
December 13	50	−9	0.35″	6.4″
December 14	61	−3	0.70″	16.4″
December 15	59	−7	1.19″	8.4″
December 16	57	−12	0.48″	5.8″
December 17	47	−17	0.92″	5.1″
December 18	51	−17	0.53″	8.0″
December 19	55	−12	0.63″	13.2″
December 20	53	−16	0.49″	6.0″
December 21	58	−11	0.48″	8.0″
December 22	59	−13	0.70″	6.9″
December 23	56	−17	0.77″	6.3″
December 24	51	−12	0.61″	4.3″
December 25	65	−18	0.63″	6.0″
December 26	50	−11	0.32″	8.5″
December 27	52	−16	1.12″	14.0″
December 28	56	−9	2.28″	9.1″
December 29	57	−27	0.95″	9.0″
December 30	51	−11	1.23″	8.1″
December 31	62	−11	0.61″	8.2″

Ann Arbor (Climate Record Begins 1880)

	RECORD HIGH	RECORD LOW	RECORD PRECIPITATION	RECORD SNOWFALL
January 1	54	−4	1.42″	11.4″
January 2	55	−8	0.97″	10.1″
January 3	60	−10	0.77″	5.8″
January 4	58	−13	1.44″	3.0″
January 5	59	−17	1.00″	4.3″
January 6	58	−12	1.62″	10.0″
January 7	63	−8	0.51″	4.0″
January 8	61	−9	1.20″	5.0″
January 9	58	−8	0.66″	6.2″
January 10	62	−11	0.78″	8.0″
January 11	63	−12	0.84″	4.7″
January 12	58	−19	1.17″	3.1″
January 13	66	−12	0.80″	8.0″
January 14	58	−10	1.54″	9.4″
January 15	61	−11	1.05″	4.0″
January 16	55	−16	0.50″	5.6″
January 17	62	−15	0.97″	5.0″
January 18	59	−10	0.75″	4.7″
January 19	61	−22	1.17″	5.3″
January 20	62	−14	1.33″	7.0″
January 21	62	−17	1.27″	5.6″
January 22	62	−19	1.27″	9.3″
January 23	59	−16	0.79″	4.8″
January 24	61	−16	0.55″	3.5″
January 25	72	−16	1.10″	6.0″
January 26	66	−8	1.65″	11.6″
January 27	63	−8	1.59″	14.0″
January 28	55	−11	1.00″	3.0″
January 29	57	−17	1.27″	7.0″
January 30	52	−11	0.90″	7.6″
January 31	59	−7	1.57″	6.9″
February 1	57	−10	1.37″	8.9″
February 2	54	−9	0.90″	5.0″
February 3	54	−12	0.98″	11.0″
February 4	62	−9	0.74″	6.5″
February 5	62	−19	1.32″	4.2″
February 6	55	−10	1.77″	4.0″
February 7	51	−9	0.75″	5.9″
February 8	63	−10	0.65″	3.9″
February 9	60	−17	2.50″	5.0″
February 10	58	−21	1.10″	7.8″
February 11	67	**−23**	1.58″	7.0″
February 12	66	−16	1.98″	5.7″
February 13	62	−16	0.90″	9.0″
February 14	58	−13	1.24″	5.8″
February 15	65	−10	1.68″	6.9″
February 16	61	−12	2.25″	5.0″

Ann Arbor *(continued)*

	RECORD HIGH	RECORD LOW	RECORD PRECIPITATION	RECORD SNOWFALL
February 17	58	–14	1.68″	5.0″
February 18	56	–9	0.88″	4.8″
February 19	59	–8	2.32″	9.0″
February 20	62	–20	1.40″	6.0″
February 21	63	–7	1.41″	5.0″
February 22	63	–9	1.88″	3.0″
February 23	65	–9	1.27″	6.1″
February 24	60	–9	0.70″	7.0″
February 25	66	–5	1.12″	6.5″
February 26	66	–4	0.73″	6.9″
February 27	64	–6	1.01″	3.7″
February 28	60	–7	1.15″	10.5″
February 29	57	–3	0.89″	3.8″
March 1	63	–5	0.74″	6.3″
March 2	62	–8	1.10″	2.8″
March 3	67	–8	0.77″	3.5″
March 4	66	–2	1.61″	5.0″
March 5	66	–1	0.80″	8.0″
March 6	68	–1	1.22″	8.0″
March 7	75	–2	0.70″	4.0″
March 8	78	–3	1.52″	7.0″
March 9	74	0	1.37″	6.0″
March 10	69	5	1.14″	8.0″
March 11	69	6	0.93″	4.0″
March 12	74	0	0.99″	3.0″
March 13	74	4	0.99″	5.0″
March 14	75	–2	1.65″	12.5″
March 15	76	3	0.96″	5.0″
March 16	70	–4	1.50″	2.8″
March 17	72	–5	1.51″	12.4″
March 18	76	5	1.08″	7.2″
March 19	77	–1	0.80″	4.5″
March 20	75	–7	1.17″	10.0″
March 21	74	–6	1.09″	8.0″
March 22	81	–2	0.95″	8.0″
March 23	72	–1	1.07″	8.0″
March 24	83	1	0.86″	8.0″
March 25	76	4	2.35″	5.0″
March 26	79	8	0.79″	2.2″
March 27	79	6	1.30″	4.3″
March 28	80	8	1.30″	4.5″
March 29	78	5	1.30″	1.3″
March 30	80	8	0.73″	6.0″
March 31	78	4	2.10″	1.6″
April 1	75	9	1.54″	5.1″
April 2	80	13	1.30″	2.8″
April 3	80	13	1.13″	3.8″
April 4	74	12	1.04″	3.2″
April 5	80	14	1.20″	1.2″
April 6	82	16	1.19″	10.0″
April 7	82	7	1.15″	4.8″
April 8	80	12	0.82″	1.8″
April 9	77	17	0.98″	2.8″
April 10	80	14	1.01″	3.8″
April 11	85	17	1.53″	3.5″
April 12	81	18	1.35″	4.5″
April 13	85	16	0.75″	0.7″
April 14	84	20	1.42″	2.5″
April 15	84	20	1.44″	1.0″
April 16	85	19	1.50″	1.5″
April 17	84	19	1.52″	6.7″
April 18	87	22	1.37″	0.4″
April 19	84	16	1.53″	0.1″
April 20	82	18	2.09″	0.7″
April 21	84	18	1.35″	0.5″
April 22	86	24	1.20″	1.2″
April 23	86	22	1.38″	0.1″
April 24	86	23	1.60″	5.5″
April 25	87	25	1.82″	0.9″
April 26	86	25	1.48″	0.5″
April 27	84	26	1.22″	0.7″
April 28	84	28	1.47″	
April 29	88	24	2.35″	2.5″
April 30	86	28	1.15″	
May 1	87	25	1.23″	
May 2	86	27	1.98″	1.2″
May 3	88	24	1.07″	0.4″
May 4	87	27	1.35″	0.2″
May 5	90	20	1.24″	
May 6	89	31	1.70″	
May 7	86	29	0.89″	
May 8	86	28	1.35″	
May 9	88	27	1.17″	2.0″
May 10	88	27	2.19″	3.5″
May 11	86	27	2.59″	
May 12	88	30	1.68″	
May 13	87	29	0.98″	0.5″
May 14	89	30	1.56″	
May 15	90	33	1.81″	
May 16	91	30	1.26″	
May 17	92	32	1.50″	
May 18	92	33	1.72″	
May 19	92	30	1.40″	
May 20	90	33	1.40″	
May 21	90	30	2.19″	
May 22	90	32	1.37″	
May 23	90	31	1.25″	
May 24	89	32	1.10″	
May 25	88	33	2.00″	
May 26	91	34	2.01″	
May 27	95	32	2.56″	

Ann Arbor *(continued)*

	RECORD HIGH	RECORD LOW	RECORD PRECIPITATION	RECORD SNOWFALL
May 28	91	32	2.23″	
May 29	92	30	1.11″	
May 30	91	35	3.00″	
May 31	91	35	2.10″	
June 1	100	37	1.52″	
June 2	98	39	1.43″	
June 3	94	37	1.35″	
June 4	93	38	1.68″	
June 5	95	38	1.54″	
June 6	94	38	2.30″	
June 7	94	40	1.39″	
June 8	96	39	1.90″	
June 9	95	38	0.76″	
June 10	94	35	1.34″	
June 11	95	38	0.80″	
June 12	95	40	1.38″	
June 13	99	44	1.41″	
June 14	95	40	1.04″	
June 15	95	43	1.53″	
June 16	95	38	1.95″	
June 17	100	44	1.33″	
June 18	94	43	1.68″	
June 19	95	37	0.76″	
June 20	97	37	2.65″	
June 21	96	40	1.82″	
June 22	94	42	1.94″	
June 23	96	39	2.93″	
June 24	94	43	1.45″	
June 25	101	41	3.17″	
June 26	98	43	3.29″	
June 27	97	46	1.90″	
June 28	103	44	1.74″	
June 29	98	44	2.91″	
June 30	97	43	1.83″	
July 1	98	45	2.21″	
July 2	102	43	1.09″	
July 3	100	41	2.43″	
July 4	100	45	0.97″	
July 5	101	45	1.23″	
July 6	100	45	1.45″	
July 7	100	37	2.45″	
July 8	102	45	1.47″	
July 9	102	46	0.95″	
July 10	101	45	1.26″	
July 11	99	45	2.69″	
July 12	103	48	1.00″	
July 13	103	48	1.55″	
July 14	103	47	1.26″	
July 15	97	45	1.90″	
July 16	99	48	2.42″	

	RECORD HIGH	RECORD LOW	RECORD PRECIPITATION	RECORD SNOWFALL
July 17	98	50	2.58″	
July 18	97	48	1.59″	
July 19	101	46	2.21″	
July 20	102	47	1.85″	
July 21	97	47	1.23″	
July 22	97	47	2.85″	
July 23	99	49	1.67″	
July 24	**105**	49	2.08″	
July 25	103	49	1.14″	
July 26	99	47	2.20″	
July 27	99	47	1.78″	
July 28	99	49	1.51″	
July 29	98	49	1.73″	
July 30	98	49	1.35″	
July 31	95	43	0.90″	
August 1	99	48	2.09″	
August 2	97	44	1.80″	
August 3	97	45	2.07″	
August 4	96	46	2.18″	
August 5	99	48	1.63″	
August 6	104	45	**4.54″**	
August 7	100	45	1.42″	
August 8	98	42	1.85″	
August 9	99	42	1.39″	
August 10	97	44	1.39″	
August 11	95	46	1.91″	
August 12	96	44	1.86″	
August 13	96	46	0.88″	
August 14	97	42	1.59″	
August 15	96	47	1.56″	
August 16	96	45	1.86″	
August 17	95	49	1.28″	
August 18	97	43	1.62″	
August 19	99	44	3.60″	
August 20	96	44	2.40″	
August 21	96	45	1.22″	
August 22	101	40	2.87″	
August 23	100	43	1.38″	
August 24	95	44	1.13″	
August 25	96	41	1.21″	
August 26	94	45	1.16″	
August 27	97	44	1.27″	
August 28	94	39	1.35″	
August 29	96	40	1.23″	
August 30	95	42	3.73″	
August 31	98	39	1.56″	
September 1	98	30	1.10″	
September 2	99	39	1.15″	
September 3	98	33	2.33″	
September 4	92	42	1.77″	

Ann Arbor *(continued)*

	RECORD HIGH	RECORD LOW	RECORD PRECIPITATION	RECORD SNOWFALL
September 5	95	42	1.33″	
September 6	95	40	2.12″	
September 7	96	40	2.66″	
September 8	98	38	1.35″	
September 9	96	31	0.99″	
September 10	95	37	1.32″	
September 11	94	35	1.61″	
September 12	94	33	1.57″	
September 13	94	33	2.86″	
September 14	96	33	2.08″	
September 15	98	36	1.21″	
September 16	96	37	1.22″	
September 17	95	34	0.81″	
September 18	93	36	1.37″	
September 19	93	34	1.50″	
September 20	91	35	1.67″	
September 21	92	33	1.15″	
September 22	92	31	1.54″	
September 23	86	33	1.19″	
September 24	89	32	1.56″	
September 25	92	34	1.82″	
September 26	90	32	1.40″	
September 27	87	31	3.94″	
September 28	90	30	0.97″	
September 29	91	29	2.12″	
September 30	89	27	1.28″	
October 1	91	21	1.84″	
October 2	88	28	1.85″	
October 3	89	28	2.60″	
October 4	89	30	2.85″	
October 5	89	29	1.59″	
October 6	89	29	2.38″	
October 7	89	29	1.89″	
October 8	88	26	0.74″	
October 9	86	27	1.29″	
October 10	87	27	1.34″	
October 11	85	28	1.51″	
October 12	83	27	1.80″	0.5″
October 13	83	27	2.07″	
October 14	85	26	1.85″	
October 15	88	28	1.55″	
October 16	85	27	2.23″	1.0″
October 17	84	23	1.41″	0.2″
October 18	83	23	2.19″	
October 19	84	23	2.14″	2.4″
October 20	80	25	1.82″	1.1″
October 21	81	19	1.37″	0.4″
October 22	80	26	2.67″	
October 23	82	23	1.38″	0.2″
October 24	81	20	1.83″	0.1″
October 25	80	23	0.61″	

	RECORD HIGH	RECORD LOW	RECORD PRECIPITATION	RECORD SNOWFALL
October 26	82	19	0.69″	0.3″
October 27	77	20	1.28″	0.8″
October 28	75	23	0.82″	2.1″
October 29	77	21	1.33″	
October 30	76	20	1.30″	1.8″
October 31	77	22	0.73″	0.6″
November 1	78	19	1.00″	1.0″
November 2	74	18	1.71″	3.0″
November 3	74	14	0.86″	7.5″
November 4	74	17	1.39″	1.0″
November 5	72	15	1.78″	7.5″
November 6	74	20	0.93″	1.4″
November 7	69	17	1.90″	7.0″
November 8	72	13	1.90″	10.0″
November 9	74	15	1.78″	5.5″
November 10	67	17	1.51″	6.0″
November 11	70	15	1.80″	3.9″
November 12	70	16	2.08″	10.0″
November 13	70	10	0.75″	10.0″
November 14	71	12	1.31″	6.0″
November 15	70	9	1.15″	5.0″
November 16	69	7	1.45″	3.5″
November 17	69	9	1.58″	2.7″
November 18	68	9	0.84″	1.2″
November 19	69	15	1.22″	3.0″
November 20	69	11	1.06″	3.7″
November 21	67	8	1.25″	1.6″
November 22	68	11	2.12″	1.5″
November 23	69	11	0.94″	4.0″
November 24	65	1	0.74″	2.1″
November 25	64	4	1.34″	5.0″
November 26	66	10	1.40″	6.0″
November 27	66	8	2.05″	6.0″
November 28	66	6	1.18″	8.2″
November 29	64	5	2.29″	3.5″
November 30	63	2	0.85″	6.0″
December 1	63	5	1.46″	**15.8**″
December 2	65	1	1.10″	4.6″
December 3	65	–2	1.02″	8.0″
December 4	63	3	0.92″	3.3″
December 5	67	2	0.89″	10.0″
December 6	67	3	1.34″	6.3″
December 7	62	–4	1.39″	9.6″
December 8	63	–7	1.73″	4.0″
December 9	62	–4	0.78″	5.4″
December 10	64	0	1.10″	7.3″
December 11	62	–5	0.82″	8.0″
December 12	63	–2	1.19″	7.0″
December 13	58	–3	1.10″	4.4″
December 14	64	–4	0.76″	6.6″

Ann Arbor *(continued)*

	RECORD HIGH	RECORD LOW	RECORD PRECIPITATION	RECORD SNOWFALL
December 15	61	–6	1.15″	6.6″
December 16	62	–6	1.22″	10.6″
December 17	60	–9	0.73″	5.3″
December 18	54	–14	0.58″	7.0″
December 19	55	–13	1.15″	7.2″
December 20	59	–3	0.86″	6.0″
December 21	60	–11	2.15″	3.3″
December 22	60	–12	0.91″	5.0″
December 23	53	–12	1.56″	7.4″
December 24	61	–20	1.10″	10.0″
December 25	61	–11	2.00″	6.9″
December 26	61	–15	0.85″	6.5″
December 27	57	–6	1.22″	4.0″
December 28	60	–7	1.10″	5.5″
December 29	62	–10	1.13″	4.2″
December 30	56	–8	1.28″	6.0″
December 31	60	–6	0.75″	7.5″

Benton Harbor (Climate Record Begins 1887)

	RECORD HIGH	RECORD LOW	RECORD PRECIPITATION	RECORD SNOWFALL
January 1	60	–10	1.06″	8.0″
January 2	60	0	1.30″	9.0″
January 3	66	–3	1.50″	10.0″
January 4	59	–3	1.37″	7.0″
January 5	62	–15	1.20″	8.0″
January 6	60	–12	0.95″	10.0″
January 7	66	–10	0.67″	4.0″
January 8	65	–13	1.80″	14.0″
January 9	61	–1	1.00″	7.0″
January 10	60	–7	1.23″	10.0″
January 11	62	–10	0.70″	5.0″
January 12	64	**–21**	2.15″	8.0″
January 13	65	–7	1.38″	7.0″
January 14	54	–5	1.31″	7.0″
January 15	60	–7	0.80″	8.0″
January 16	60	–17	0.65″	8.3″
January 17	59	–7	0.52″	8.0″
January 18	57	–13	0.74″	13.0″
January 19	62	–17	1.64″	8.0″
January 20	64	–15	1.03″	6.0″
January 21	66	–13	1.43″	8.0″
January 22	57	–9	1.45″	12.0″
January 23	62	–15	0.80″	8.0″
January 24	64	–15	1.00″	15.0″
January 25	68	–8	0.54″	6.0″
January 26	66	–6	1.80″	22.0″
January 27	62	–7	0.85″	20.0″
January 28	57	–7	1.94″	14.0″
January 29	58	–8	1.42″	12.0″
January 30	60	–8	0.96″	13.0″
January 31	62	–3	1.20″	12.0″
February 1	64	–9	1.20″	9.0″
February 2	50	–12	0.51″	6.0″
February 3	56	–10	0.96″	5.0″
February 4	66	–6	0.57″	7.0″
February 5	54	–9	1.12″	7.0″
February 6	55	–6	1.23″	12.0″
February 7	59	–13	0.48″	12.0″
February 8	63	–8	1.24″	8.0″
February 9	60	–12	1.13″	6.0″
February 10	62	–12	1.70″	5.0″
February 11	70	–11	1.25″	5.0″
February 12	70	–12	1.20″	14.0″
February 13	60	–12	1.40″	7.0″
February 14	60	0	1.08″	8.0″
February 15	65	–5	0.87″	11.0″
February 16	64	–8	1.20″	5.3″
February 17	62	–8	1.08″	10.0″
February 18	63	–6	1.15″	8.5″
February 19	63	–8	1.27″	4.0″
February 20	63	–5	0.50″	5.0″
February 21	65	–4	2.37″	24.0″
February 22	65	–7	0.75″	7.0″
February 23	65	–1	1.40″	9.0″
February 24	64	–8	1.00″	10.0″
February 25	69	–2	1.50″	7.0″
February 26	65	–5	0.90″	5.0″
February 27	71	–6	0.90″	4.0″
February 28	63	0	1.40″	14.0″
February 29	61	3	0.10″	1.0″
March 1	63	–2	1.30″	10.0″
March 2	65	–1	1.46″	6.0″
March 3	75	2	0.80″	5.0″
March 4	74	6	1.68″	6.0″
March 5	76	–6	1.00″	6.0″
March 6	75	–4	0.96″	8.0″
March 7	75	–3	1.10″	10.0″
March 8	78	0	1.75″	8.0″
March 9	78	2	1.43″	8.0″
March 10	70	7	1.36″	5.0″
March 11	75	3	1.20″	5.0″
March 12	77	–6	1.50″	3.0″

Benton Harbor *(continued)*

	RECORD HIGH	RECORD LOW	RECORD PRECIPITATION	RECORD SNOWFALL
March 13	80	9	1.95″	3.0″
March 14	75	5	1.20″	12.0″
March 15	73	4	1.73″	7.0″
March 16	75	5	1.05″	4.0″
March 17	77	1	1.40″	6.0″
March 18	73	5	0.73″	3.0″
March 19	75	5	2.50″	4.0″
March 20	71	7	0.76″	4.0″
March 21	76	7	1.30″	7.0″
March 22	78	7	1.04″	11.0″
March 23	80	11	0.80″	7.0″
March 24	72	5	1.00″	5.0″
March 25	79	8	1.45″	12.0″
March 26	80	5	1.66″	15.0″
March 27	81	6	1.06″	5.2″
March 28	78	12	2.20″	6.5″
March 29	81	13	1.70″	3.5″
March 30	80	11	1.24″	1.7″
March 31	84	9	2.09″	3.0″
April 1	83	12	0.90″	3.0″
April 2	79	20	1.38″	5.8″
April 3	80	18	1.77″	3.5″
April 4	75	15	1.40″	6.0″
April 5	82	16	4.20″	6.5″
April 6	82	13	1.95″	6.5″
April 7	82	9	0.88″	3.0″
April 8	77	15	1.00″	3.5″
April 9	74	15	1.26″	1.5″
April 10	80	15	1.00″	1.0″
April 11	80	18	1.04″	4.0″
April 12	82	21	1.53″	0.5″
April 13	83	18	0.95″	3.5″
April 14	82	21	1.05″	1.0″
April 15	88	16	1.20″	1.0″
April 16	89	24	1.45″	4.0″
April 17	88	25	2.40″	6.0″
April 18	88	19	2.28″	1.0″
April 19	87	18	1.70″	
April 20	84	18	1.88″	
April 21	86	23	1.23″	1.0″
April 22	86	21	1.67″	
April 23	85	18	1.62″	3.0″
April 24	85	25	4.00″	2.0″
April 25	87	24	3.29″	1.5″
April 26	88	17	0.90″	
April 27	87	25	1.69″	1.5″
April 28	85	24	2.25″	
April 29	90	27	2.40″	
April 30	88	22	2.00″	
May 1	87	27	2.00″	
May 2	87	25	1.87″	
May 3	87	29	1.77″	
May 4	90	25	1.59″	
May 5	88	30	1.40″	
May 6	89	25	1.60″	
May 7	85	26	2.00″	
May 8	87	25	1.30″	
May 9	89	25	0.86″	1.0″
May 10	90	23	1.91″	
May 11	87	28	1.75″	
May 12	89	30	1.89″	
May 13	88	26	1.75″	
May 14	92	32	1.83″	
May 15	90	31	1.55″	
May 16	89	28	2.45″	
May 17	93	30	1.20″	
May 18	92	32	1.47″	
May 19	93	30	2.70″	
May 20	91	32	0.80″	
May 21	91	28	1.48″	
May 22	90	32	1.90″	
May 23	90	30	2.00″	
May 24	91	34	1.96″	
May 25	90	27	1.77″	
May 26	90	26	1.50″	
May 27	92	24	1.90″	
May 28	93	26	1.50″	
May 29	97	32	1.08″	
May 30	95	32	**6.60″**	
May 31	95	31	1.75″	
June 1	**104**	31	2.10″	
June 2	97	33	3.10″	
June 3	93	36	1.32″	
June 4	93	35	2.00″	
June 5	95	37	1.35″	
June 6	92	34	1.75″	
June 7	96	35	1.04″	
June 8	94	36	2.09″	
June 9	92	37	1.04″	
June 10	94	36	2.01″	
June 11	94	34	1.30″	
June 12	94	42	1.82″	
June 13	96	38	2.06″	
June 14	93	41	1.03″	
June 15	94	40	1.50″	
June 16	93	41	0.82″	
June 17	95	39	3.51″	
June 18	94	40	1.65″	
June 19	99	43	1.05″	
June 20	101	42	1.25″	
June 21	98	39	3.10″	

Benton Harbor *(continued)*

	RECORD HIGH	RECORD LOW	RECORD PRECIPITATION	RECORD SNOWFALL
June 22	97	32	1.78″	
June 23	98	43	2.47″	
June 24	100	41	2.30″	
June 25	99	39	2.37″	
June 26	99	41	1.35″	
June 27	97	44	1.52″	
June 28	96	35	2.85″	
June 29	96	45	2.09″	
June 30	98	38	1.56″	
July 1	97	39	1.68″	
July 2	97	37	1.20″	
July 3	99	39	2.18″	
July 4	100	40	1.80″	
July 5	99	41	2.04″	
July 6	98	38	1.20″	
July 7	98	43	2.65″	
July 8	98	39	0.90″	
July 9	97	46	3.14″	
July 10	100	45	1.63″	
July 11	100	41	4.85″	
July 12	95	48	1.40″	
July 13	99	48	1.02″	
July 14	100	46	2.40″	
July 15	99	47	1.60″	
July 16	100	41	2.03″	
July 17	98	45	1.98″	
July 18	99	48	1.80″	
July 19	96	47	1.75″	
July 20	96	43	1.85″	
July 21	**104**	43	1.62″	
July 22	103	43	2.04″	
July 23	98	40	2.51″	
July 24	100	42	2.50″	
July 25	99	45	1.80″	
July 26	97	49	2.07″	
July 27	97	45	1.25″	
July 28	99	48	2.00″	
July 29	100	46	1.22″	
July 30	**104**	39	1.38″	
July 31	101	43	1.50″	
August 1	101	41	2.66″	
August 2	99	38	1.53″	
August 3	99	45	2.00″	
August 4	100	41	1.45″	
August 5	98	42	2.05″	
August 6	98	44	2.40″	
August 7	96	39	2.10″	
August 8	97	40	1.65″	
August 9	100	41	2.05″	
August 10	98	46	2.16″	
August 11	97	45	1.33″	
August 12	95	43	3.00″	
August 13	95	43	1.15″	
August 14	96	39	3.25″	
August 15	97	38	1.26″	
August 16	97	43	2.21″	
August 17	100	40	2.32″	
August 18	97	42	1.26″	
August 19	95	42	2.18″	
August 20	97	39	2.45″	
August 21	98	43	1.83″	
August 22	95	40	1.80″	
August 23	95	42	1.40″	
August 24	96	41	1.83″	
August 25	93	39	2.07″	
August 26	96	36	1.48″	
August 27	94	45	1.70″	
August 28	99	37	2.00″	
August 29	94	38	1.50″	
August 30	93	40	3.00″	
August 31	95	44	0.97″	
September 1	98	40	2.75″	
September 2	98	41	2.00″	
September 3	95	37	1.53″	
September 4	91	37	1.04″	
September 5	93	40	2.42″	
September 6	96	37	1.12″	
September 7	96	36	1.26″	
September 8	96	34	2.00″	
September 9	98	39	1.58″	
September 10	97	39	2.06″	
September 11	94	34	2.04″	
September 12	93	33	5.51″	
September 13	95	32	2.20″	
September 14	94	33	2.25″	
September 15	94	34	1.22″	
September 16	95	35	2.09″	
September 17	91	32	1.85″	
September 18	90	35	1.88″	
September 19	90	37	1.70″	
September 20	92	36	2.00″	
September 21	90	30	1.62″	
September 22	92	34	1.64″	
September 23	91	30	1.90″	
September 24	93	28	2.06″	
September 25	91	29	1.60″	
September 26	92	28	2.00″	
September 27	92	23	1.45″	
September 28	86	24	1.75″	
September 29	96	29	2.09″	
September 30	93	27	1.75″	0.2″

Benton Harbor *(continued)*

	RECORD HIGH	RECORD LOW	RECORD PRECIPITATION	RECORD SNOWFALL
October 1	92	26	2.20″	
October 2	87	28	1.42″	
October 3	90	28	2.44″	
October 4	94	27	1.34″	
October 5	94	25	1.72″	
October 6	88	27	1.85″	
October 7	90	26	1.65″	
October 8	87	23	1.25″	
October 9	82	19	1.45″	
October 10	85	23	1.80″	3.0″
October 11	85	20	1.87″	2.0″
October 12	85	23	2.60″	1.0″
October 13	85	20	2.85″	
October 14	84	24	1.62″	1.7″
October 15	85	26	0.92″	
October 16	85	24	1.50″	
October 17	84	24	1.70″	
October 18	82	25	1.73″	1.5″
October 19	82	23	2.81″	4.0″
October 20	86	23	1.44″	2.0″
October 21	85	21	1.80″	
October 22	85	21	0.91″	
October 23	83	20	2.20″	4.0″
October 24	80	22	1.80″	
October 25	80	22	1.13″	
October 26	80	19	1.30″	
October 27	80	21	2.30″	2.5″
October 28	76	24	0.75″	1.0″
October 29	78	18	1.10″	2.0″
October 30	81	15	2.88″	6.5″
October 31	82	15	1.15″	5.0″
November 1	82	21	1.61″	4.0″
November 2	77	17	1.45″	
November 3	76	15	1.10″	4.0″
November 4	76	14	1.48″	5.0″
November 5	76	14	2.78″	3.0″
November 6	77	16	0.97″	4.5″
November 7	76	17	1.21″	2.2″
November 8	72	19	1.92″	2.0″
November 9	72	13	1.50″	2.5″
November 10	71	15	1.75″	2.0″
November 11	74	12	1.30″	4.0″
November 12	72	20	1.02″	3.0″
November 13	73	14	1.28″	2.0″
November 14	71	10	1.65″	4.0″
November 15	71	12	1.40″	2.0″

	RECORD HIGH	RECORD LOW	RECORD PRECIPITATION	RECORD SNOWFALL
November 16	68	10	0.83″	7.0″
November 17	73	11	1.45″	3.0″
November 18	72	12	0.83″	7.0″
November 19	76	8	2.00″	3.8″
November 20	71	18	1.67″	7.0″
November 21	68	11	0.89″	12.0″
November 22	68	9	1.32″	8.0″
November 23	67	11	1.04″	3.0″
November 24	65	–1	0.77″	3.0″
November 25	68	–19	1.50″	8.0″
November 26	64	13	1.70″	7.0″
November 27	66	10	2.74″	8.0″
November 28	67	7	1.82″	7.0″
November 29	67	6	1.44″	4.0″
November 30	67	11	1.27″	10.0″
December 1	69	10	0.80″	7.0″
December 2	69	1	0.70″	3.0″
December 3	67	–2	2.37″	6.0″
December 4	64	4	1.02″	8.0″
December 5	68	5	1.50″	5.0″
December 6	67	6	1.31″	**25.0″**
December 7	62	7	1.04″	5.0″
December 8	62	–3	1.00″	10.0″
December 9	62	3	1.27″	8.0″
December 10	64	4	1.50″	8.1″
December 11	63	0	0.88″	12.0″
December 12	62	–4	1.80″	13.0″
December 13	65	–5	1.07″	6.0″
December 14	65	0	1.10″	8.0″
December 15	62	–7	1.42″	7.0″
December 16	64	–6	1.75″	12.0″
December 17	59	–8	0.95″	5.0″
December 18	54	–3	1.52″	8.0″
December 19	58	–2	1.94″	10.0″
December 20	58	–9	1.80″	7.0″
December 21	60	–9	2.10″	10.0″
December 22	58	–5	1.11″	10.0″
December 23	59	–15	1.30″	8.0″
December 24	65	–7	1.16″	6.5″
December 25	66	–5	1.40″	7.0″
December 26	58	–3	0.72″	6.0″
December 27	61	–2	1.50″	5.0″
December 28	65	–4	0.75″	6.0″
December 29	64	–4	0.87″	6.0″
December 30	68	–7	1.52″	4.0″
December 31	61	–9	1.50″	15.0″

Big Rapids (Climate Record Begins 1896)

	RECORD HIGH	RECORD LOW	RECORD PRECIPITATION	RECORD SNOWFALL
January 1	53	−19	0.73″	6.5″
January 2	54	−5	1.06″	7.3″
January 3	55	−22	2.20″	6.0″
January 4	52	−23	1.15″	5.0″
January 5	49	−16	0.60″	6.0″
January 6	53	−13	0.65″	6.3″
January 7	52	−16	1.80″	8.0″
January 8	58	−19	1.05″	4.0″
January 9	52	−13	1.27″	5.6″
January 10	54	−17	0.96″	4.0″
January 11	51	−22	0.61″	8.0″
January 12	44	−24	1.63″	6.7″
January 13	52	−18	0.96″	10.5″
January 14	49	−12	1.02″	6.2″
January 15	51	−25	1.10″	7.5″
January 16	54	−23	0.42″	7.0″
January 17	51	−16	2.07″	3.5″
January 18	56	−28	0.71″	9.0″
January 19	51	−25	1.80″	6.5″
January 20	55	−22	0.98″	10.0″
January 21	54	−25	1.73″	8.0″
January 22	51	−17	1.14″	8.0″
January 23	51	−17	1.26″	12.0″
January 24	52	−22	1.06″	4.2″
January 25	55	−25	1.43″	4.4″
January 26	61	−21	1.55″	14.0″
January 27	57	−17	0.49″	9.4″
January 28	47	−18	0.84″	10.0″
January 29	52	−18	0.94″	6.0″
January 30	50	−30	0.81″	**16.0**″
January 31	51	−23	0.91″	6.0″
February 1	47	−27	0.97″	8.0″
February 2	48	−20	0.51″	5.0″
February 3	51	−29	0.41″	5.1″
February 4	51	−26	0.89″	5.0″
February 5	50	−26	1.24″	6.0″
February 6	54	−18	0.97″	10.6″
February 7	47	−25	0.89″	5.0″
February 8	56	−24	1.30″	4.0″
February 9	53	−23	1.04″	5.0″
February 10	52	−33	0.78″	8.6″
February 11	63	**−36**	1.62″	6.6″
February 12	52	−34	0.80″	3.5″
February 13	54	−28	1.34″	5.2″
February 14	50	−21	1.17″	4.0″
February 15	57	−23	0.62″	4.0″
February 16	56	−26	0.51″	3.8″
February 17	49	−26	1.00″	6.0″
February 18	55	−14	0.50″	6.3″
February 19	56	−19	1.40″	6.0″
February 20	59	−24	0.98″	7.1″
February 21	61	−19	1.44″	6.3″
February 22	62	−22	2.51″	8.0″
February 23	60	−10	0.79″	9.1″
February 24	57	−18	0.77″	6.0″
February 25	56	−18	1.40″	7.1″
February 26	64	−15	1.66″	12.0″
February 27	52	−17	0.83″	6.0″
February 28	54	−7	0.65″	8.3″
February 29	61	−16	0.52″	4.6″
March 1	64	−14	0.70″	3.5″
March 2	57	−15	0.95″	10.0″
March 3	67	−24	0.92″	10.4″
March 4	59	−9	1.77″	10.2″
March 5	62	−9	0.58″	4.3″
March 6	65	−15	1.45″	6.0″
March 7	71	−8	1.00″	4.6″
March 8	76	−16	0.85″	8.6″
March 9	68	−7	1.21″	8.0″
March 10	64	−9	1.06″	5.0″
March 11	67	−8	1.22″	6.0″
March 12	73	−14	1.34″	5.5″
March 13	74	−9	1.09″	9.0″
March 14	75	0	1.36″	4.5″
March 15	76	−3	1.14″	5.3″
March 16	69	−11	1.30″	3.0″
March 17	72	−5	1.68″	8.5″
March 18	72	−3	1.09″	7.0″
March 19	74	−5	1.12″	7.0″
March 20	72	−4	1.02″	4.0″
March 21	72	−3	1.05″	7.0″
March 22	77	−4	1.77″	4.0″
March 23	76	−7	0.87″	4.0″
March 24	75	−4	0.88″	6.0″
March 25	79	−5	1.02″	5.0″
March 26	76	0	1.10″	9.5″
March 27	79	−1	2.22″	4.5″
March 28	79	4	1.28″	4.0″
March 29	82	3	0.86″	5.0″
March 30	74	5	0.75″	5.0″
March 31	76	0	1.36″	7.5″
April 1	72	5	1.24″	3.1″
April 2	79	12	0.91″	7.0″
April 3	80	7	1.03″	4.0″
April 4	78	9	1.01″	4.0″
April 5	80	8	1.33″	3.0″
April 6	80	3	1.20″	3.0″
April 7	82	1	2.45″	7.5″
April 8	72	7	1.58″	5.0″

Big Rapids *(continued)*

	RECORD HIGH	RECORD LOW	RECORD PRECIPITATION	RECORD SNOWFALL
April 9	77	11	0.65″	5.0″
April 10	79	13	1.35″	3.0″
April 11	80	10	1.25″	2.3″
April 12	83	14	1.00″	4.1″
April 13	80	15	1.19″	0.5″
April 14	79	15	1.77″	2.0″
April 15	84	12	1.40″	14.0″
April 16	87	10	1.70″	2.8″
April 17	81	20	1.67″	3.5″
April 18	84	20	0.96″	2.0″
April 19	89	16	2.47″	
April 20	81	13	1.03″	
April 21	82	17	1.32″	1.1″
April 22	86	21	1.11″	
April 23	85	18	1.15″	1.2″
April 24	86	21	2.32″	0.1″
April 25	84	16	1.96″	
April 26	86	20	1.40″	
April 27	83	22	1.60″	0.1″
April 28	83	23	1.18″	0.1″
April 29	87	23	1.20″	12.0″
April 30	84	24	1.75″	3.0″
May 1	86	20	1.26″	
May 2	88	22	1.30″	
May 3	87	22	1.16″	3.0″
May 4	89	20	1.23″	
May 5	89	20	1.10″	
May 6	87	23	1.40″	0.3″
May 7	85	22	1.01″	
May 8	87	19	1.31″	3.5″
May 9	85	23	1.46″	1.0″
May 10	88	22	1.78″	
May 11	87	24	2.62″	
May 12	89	23	1.07″	
May 13	89	23	1.36″	
May 14	88	23	1.43″	
May 15	89	26	1.20″	
May 16	88	26	1.89″	
May 17	89	27	1.56″	
May 18	90	28	1.19″	
May 19	91	26	1.69″	
May 20	92	26	2.50″	
May 21	89	23	1.78″	
May 22	88	26	1.93″	
May 23	87	25	1.70″	
May 24	89	29	0.88″	
May 25	88	29	1.30″	
May 26	92	29	1.45″	
May 27	91	27	1.16″	3.0″
May 28	90	27	0.80″	
May 29	90	31	1.03″	

	RECORD HIGH	RECORD LOW	RECORD PRECIPITATION	RECORD SNOWFALL
May 30	93	30	1.43″	
May 31	90	32	1.07″	
June 1	97	31	2.05″	
June 2	90	33	2.33″	
June 3	89	32	0.85″	
June 4	90	31	1.86″	
June 5	93	32	2.05″	
June 6	93	32	2.22″	
June 7	94	35	1.04″	
June 8	93	30	1.34″	
June 9	92	28	1.78″	
June 10	93	30	1.83″	
June 11	91	31	1.80″	
June 12	94	35	4.00″	
June 13	93	35	2.20″	
June 14	94	34	2.08″	
June 15	93	35	1.85″	
June 16	95	33	0.96″	
June 17	95	32	3.24″	
June 18	96	37	1.55″	
June 19	98	38	2.15″	
June 20	97	37	1.38″	
June 21	96	34	0.97″	
June 22	93	34	1.42″	
June 23	94	35	1.21″	
June 24	93	39	1.72″	
June 25	95	38	2.31″	
June 26	96	40	3.05″	
June 27	99	39	1.18″	
June 28	98	42	1.30″	
June 29	96	40	1.77″	
June 30	98	37	1.70″	
July 1	97	38	1.50″	
July 2	97	37	1.41″	
July 3	98	37	1.48″	
July 4	98	38	3.48″	
July 5	99	40	1.20″	
July 6	100	38	1.40″	
July 7	99	42	2.42″	
July 8	100	41	1.31″	
July 9	98	41	1.10″	
July 10	101	38	0.76″	
July 11	100	32	1.58″	
July 12	102	40	1.63″	
July 13	**103**	42	2.04″	
July 14	**103**	39	1.78″	
July 15	95	42	1.21″	
July 16	97	40	1.22″	
July 17	95	43	2.33″	
July 18	95	42	3.94″	

Big Rapids *(continued)*

	RECORD HIGH	RECORD LOW	RECORD PRECIPITATION	RECORD SNOWFALL
July 19	95	40	1.38″	
July 20	97	39	1.57″	
July 21	95	41	1.55″	
July 22	96	43	1.78″	
July 23	98	38	1.75″	
July 24	99	41	3.62″	
July 25	98	43	1.67″	
July 26	99	40	2.28″	
July 27	97	41	2.53″	
July 28	99	44	1.55″	
July 29	102	41	1.50″	
July 30	**103**	41	1.63″	
July 31	95	43	1.62″	
August 1	99	39	1.68″	
August 2	96	37	2.25″	
August 3	99	38	1.93″	
August 4	98	39	3.54″	
August 5	99	42	2.80″	
August 6	100	42	1.53″	
August 7	99	38	0.92″	
August 8	99	37	1.76″	
August 9	94	37	1.41″	
August 10	96	41	1.44″	
August 11	94	39	1.18″	
August 12	95	38	3.10″	
August 13	92	39	1.78″	
August 14	93	39	1.46″	
August 15	96	39	0.99″	
August 16	95	35	1.17″	
August 17	101	37	1.36″	
August 18	94	37	1.00″	
August 19	97	37	2.00″	
August 20	99	34	1.56″	
August 21	99	36	1.41″	
August 22	94	38	1.24″	
August 23	97	36	1.44″	
August 24	97	37	3.53″	
August 25	95	35	1.56″	
August 26	94	37	2.90″	
August 27	97	34	2.68″	
August 28	94	36	1.35″	
August 29	94	32	1.91″	
August 30	94	32	1.60″	
August 31	93	36	4.55″	
September 1	97	33	2.04″	
September 2	97	30	1.45″	
September 3	92	31	1.42″	
September 4	90	33	2.67″	
September 5	90	35	2.15″	
September 6	90	31	1.97″	
September 7	94	32	2.22″	
September 8	93	33	1.04″	
September 9	92	34	2.52″	
September 10	92	32	5.44″	
September 11	93	29	**7.64**″	
September 12	94	31	2.30″	
September 13	93	32	2.40″	
September 14	95	25	3.30″	
September 15	94	30	1.00″	
September 16	87	30	1.04″	
September 17	89	30	1.69″	
September 18	90	29	1.39″	
September 19	89	29	1.56″	
September 20	87	30	1.55″	
September 21	90	26	2.10″	
September 22	89	22	1.39″	
September 23	85	16	1.50″	
September 24	85	25	1.55″	
September 25	88	27	1.93″	
September 26	86	26	1.41″	
September 27	84	23	1.77″	
September 28	87	24	1.98″	
September 29	88	26	1.61″	
September 30	86	26	1.99″	
October 1	87	19	1.55″	
October 2	86	24	0.64″	
October 3	83	23	3.07″	
October 4	84	23	2.46″	
October 5	83	22	0.87″	
October 6	84	22	2.73″	
October 7	86	20	1.90″	7.0″
October 8	87	19	1.51″	
October 9	86	19	0.80″	
October 10	85	21	1.50″	
October 11	84	20	0.89″	
October 12	82	21	2.00″	0.5″
October 13	84	20	1.18″	
October 14	82	22	1.35″	2.0″
October 15	88	21	1.39″	
October 16	84	20	1.32″	
October 17	85	19	1.25″	2.0″
October 18	81	18	1.62″	
October 19	79	21	0.88″	1.0″
October 20	82	20	1.26″	
October 21	83	18	1.77″	1.1″
October 22	80	19	1.38″	0.5″
October 23	81	19	2.40″	0.5″
October 24	79	19	1.74″	2.0″
October 25	76	19	1.11″	5.4″
October 26	78	10	1.25″	4.0″
October 27	78	17	1.13″	5.0″

Big Rapids *(continued)*

	RECORD HIGH	RECORD LOW	RECORD PRECIPITATION	RECORD SNOWFALL
October 28	72	18	0.92″	2.5″
October 29	75	16	1.14″	3.0″
October 30	77	16	2.00″	3.0″
October 31	76	19	1.77″	0.2″
November 1	76	20	1.65″	0.6″
November 2	75	17	1.42″	2.0″
November 3	74	10	1.47″	8.0″
November 4	70	11	1,67″	2.0″
November 5	71	6	1.90″	6.0″
November 6	72	5	1.61″	0.5″
November 7	70	11	1.40″	8.0″
November 8	68	9	1.09″	4.0″
November 9	73	11	1.21″	11.7″
November 10	67	17	1.61″	2.0″
November 11	67	9	0.97″	3.7″
November 12	65	9	1.25″	2.5″
November 13	69	6	1.20″	3.8″
November 14	65	4	1.53″	2.0″
November 15	66	–1	2.68″	2.5″
November 16	65	7	1.11″	2.5″
November 17	68	5	2.10″	4.4″
November 18	68	11	1.32″	2.0″
November 19	69	10	1.35″	3.0″
November 20	65	7	1.25″	9.0″
November 21	63	8	1.44″	1.5″
November 22	62	7	1.04″	6.0″
November 23	64	–2	0.92″	4.5″
November 24	64	–7	0.84″	4.0″
November 25	58	–8	0.88″	4.0″
November 26	62	0	1.11″	7.0″
November 27	64	1	2.18″	9.2″
November 28	61	1	1.05″	3.5″
November 29	64	–1	1.15″	4.0″
November 30	64	–2	1.15″	8.0″
December 1	63	–2	0.88″	4.5″
December 2	62	–7	2.13″	7.0″
December 3	60	–9	1.42″	5.5″
December 4	59	–7	0.55″	3.3″
December 5	66	–3	0.93″	5.2″
December 6	59	–6	0.92″	8.0″
December 7	59	–7	1.00″	10.0″
December 8	56	–2	1.33″	8.0″
December 9	59	–4	0.60″	5.5″
December 10	60	–8	1.95″	10.5″
December 11	57	–5	1.40″	12.0″
December 12	59	–8	0.91″	6.0″
December 13	55	–16	1.12″	4.0″
December 14	58	–8	0.60″	4.0″
December 15	51	–15	1.68″	11.4″
December 16	56	–9	0.82″	6.0″
December 17	49	–5	0.90″	9.0″
December 18	51	–16	0.96″	7.0″
December 19	50	–15	1.00″	8.0″
December 20	55	–12	0.91″	8.0″
December 21	57	–10	0.95″	9.0″
December 22	54	–11	0.90″	4.0″
December 23	54	–15	0.65″	4.5″
December 24	51	–12	1.03″	8.5″
December 25	61	–11	0.56″	5.0″
December 26	47	–11	0.56″	3.0″
December 27	51	–9	1.94″	6.0″
December 28	60	–13	1.76″	5.2″
December 29	59	–15	0.65″	5.5″
December 30	56	–16	1.31″	10.0″
December 31	57	–18	0.74″	11.0″

Bloomingdale (Climate Record Begins 1904)

	RECORD HIGH	RECORD LOW	RECORD PRECIPITATION	RECORD SNOWFALL
January 1	58	–14	1.89″	6.0″
January 2	54	–6	1.10″	8.0″
January 3	61	–12	0.69″	10.0″
January 4	55	–17	1.39″	12.0″
January 5	62	–18	0.98″	9.5″
January 6	55	–10	1.00″	8.0″
January 7	59	–6	1.00″	6.0″
January 8	65	–13	1.75″	5.0″
January 9	57	–10	0.95″	6.5″
January 10	54	–10	0.67″	10.7″
January 11	58	–15	0.92″	6.0″
January 12	55	–18	2.55″	8.0″
January 13	62	–12	1.10″	6.6″
January 14	57	–8	0.98″	11.0″
January 15	56	–13	0.92″	7.0″
January 16	56	–16	0.49″	8.0″
January 17	54	–14	0.45″	5.0″
January 18	55	–14	0.94″	5.0″
January 19	62	–15	1.50″	8.0″
January 20	63	–17	0.77″	7.0″

Bloomingdale *(continued)*

	RECORD HIGH	RECORD LOW	RECORD PRECIPITATION	RECORD SNOWFALL
January 21	61	–13	1.10″	7.0″
January 22	57	–14	1.45″	8.0″
January 23	60	–12	0.57″	6.0″
January 24	57	–16	0.60″	6.0″
January 25	67	–14	0.50″	5.0″
January 26	66	–12	0.94″	10.0″
January 27	62	–12	1.00″	10.0″
January 28	63	–12	0.89″	14.0″
January 29	59	–5	0.96″	8.0″
January 30	49	–13	1.25″	7.0″
January 31	55	–7	1.16″	10.0″
February 1	64	–16	1.00″	6.0″
February 2	53	–8	0.77″	5.5″
February 3	51	**–23**	0.70″	5.0″
February 4	57	–21	0.60″	8.0″
February 5	54	–22	0.81″	8.0″
February 6	55	–11	1.00″	7.0″
February 7	54	–22	0.73″	7.0″
February 8	58	–21	0.78″	7.0″
February 9	58	–17	0.57″	7.0″
February 10	52	–22	1.65″	8.0″
February 11	63	–15	0.60″	5.0″
February 12	71	–17	1.09″	7.0″
February 13	61	–14	0.90″	7.1″
February 14	55	–16	1.50″	9.5″
February 15	64	–17	0.78″	9.0″
February 16	63	–11	1.42″	4.0″
February 17	57	–13	0.90″	5.0″
February 18	60	–14	1.03″	5.0″
February 19	61	–15	1.13″	8.0″
February 20	65	–15	0.69″	4.0″
February 21	67	–12	2.21″	4.0″
February 22	65	–16	1.60″	10.0″
February 23	62	–7	1.21″	6.0″
February 24	65	–16	0.70″	5.0″
February 25	63	–3	1.55″	7.0″
February 26	66	–12	0.60″	5.0″
February 27	57	–13	2.00″	10.0″
February 28	69	–13	1.10″	3.5″
February 29	56	–1	0.60″	12.0″
March 1	62	–2	1.08″	4.0″
March 2	65	–9	0.93″	8.0″
March 3	63	–10	0.94″	9.0″
March 4	73	–4	1.48″	7.0″
March 5	70	–5	1.10″	6.0″
March 6	74	–8	1.50″	8.0″
March 7	70	–3	0.87″	4.0″
March 8	76	–8	2.25″	6.0″
March 9	79	–6	1.11″	7.5″
March 10	65	–4	0.91″	9.0″
March 11	65	0	1.03″	3.0″
March 12	74	–2	0.65″	5.0″
March 13	79	–1	1.11″	3.5″
March 14	74	–5	3.90″	4.0″
March 15	77	–1	1.47″	2.0″
March 16	74	–2	1.43″	3.0″
March 17	70	–1	0.80″	5.0″
March 18	73	6	0.70″	2.5″
March 19	76	–1	2.55″	3.1″
March 20	75	5	0.70″	3.5″
March 21	77	9	0.78″	7.5″
March 22	82	8	1.40″	12.0″
March 23	82	–1	0.80″	7.0″
March 24	81	–1	1.80″	2.0″
March 25	77	–2	1.82″	5.0″
March 26	81	4	1.00″	10.0″
March 27	80	1	0.79″	3.0″
March 28	79	2	2.25″	1.3″
March 29	79	8	0.91″	10.0″
March 30	79	6	1.37″	3.0″
March 31	78	3	1.21″	3.5″
April 1	82	9	1.20″	7.0″
April 2	81	19	1.66″	4.0″
April 3	82	12	1.25″	8.0″
April 4	78	13	1.45″	2.0″
April 5	79	15	4.00″	6.0″
April 6	81	12	1.42″	5.0″
April 7	82	6	1.43″	5.0″
April 8	81	5	1.27″	6.0″
April 9	80	11	1.23″	2.0″
April 10	85	14	1.10″	4.0″
April 11	81	12	1.35″	3.0″
April 12	82	19	1.11″	6.0″
April 13	85	19	0.97″	5.0″
April 14	82	21	0.98″	1.2″
April 15	82	16	2.10″	5.0″
April 16	89	21	1.41″	10.0″
April 17	88	21	1.52″	7.0″
April 18	85	19	0.89″	0.5″
April 19	89	16	4.70″	4.0″
April 20	84	15	2.39″	2.0″
April 21	84	21	1.44″	1.0″
April 22	86	21	1.30″	0.5″
April 23	87	20	1.77″	1.0″
April 24	85	20	1.77″	1.0″
April 25	85	21	1.32″	
April 26	87	23	1.84″	2.0″
April 27	87	23	1.02″	3.0″
April 28	84	26	2.10″	
April 29	85	26	1.48″	
April 30	89	20	1.05″	2.0″

Bloomingdale *(continued)*

	RECORD HIGH	RECORD LOW	RECORD PRECIPITATION	RECORD SNOWFALL
May 1	88	20	1.50″	0.8″
May 2	90	22	1.57″	3.0″
May 3	89	25	1.35″	
May 4	90	23	1.10″	0.5″
May 5	92	26	1.39″	
May 6	90	24	0.97″	
May 7	88	24	1.52″	
May 8	89	27	0.93″	
May 9	87	24	1.16″	2.0″
May 10	88	24	1.42″	
May 11	88	24	2.86″	
May 12	86	27	2.29″	
May 13	86	24	2.67″	
May 14	87	14	1.80″	
May 15	89	18	1.45″	
May 16	91	27	2.30″	
May 17	91	27	1.15″	
May 18	93	27	1.29″	
May 19	91	30	3.77″	
May 20	92	30	0.88″	
May 21	91	28	1.64″	
May 22	90	29	1.18″	
May 23	90	30	1.66″	
May 24	89	30	1.58″	
May 25	90	29	1.95″	
May 26	91	32	0.89″	
May 27	94	29	1.65″	
May 28	93	32	1.55″	
May 29	94	32	1.24″	
May 30	94	31	1.10″	
May 31	92	32	2.20″	
June 1	99	33	1.59″	
June 2	94	32	1.91″	
June 3	95	33	1.60″	
June 4	93	34	2.30″	
June 5	94	34	1.08″	
June 6	94	32	1.86″	
June 7	94	36	0.80″	
June 8	94	34	1.87″	
June 9	93	35	1.57″	
June 10	94	34	2.05″	
June 11	94	31	1.44″	
June 12	95	39	1.11″	
June 13	94	35	2.01″	
June 14	92	35	1.08″	
June 15	95	40	2.42″	
June 16	92	35	1.63″	
June 17	95	39	1.96″	
June 18	97	36	3.20″	
June 19	98	41	1.65″	
June 20	97	40	1.40″	
June 21	97	39	4.40″	
June 22	98	33	1.67″	
June 23	98	33	1.37″	
June 24	93	39	2.98″	
June 25	96	39	1.81″	
June 26	98	42	3.57″	
June 27	97	42	2.93″	
June 28	99	38	3.84″	
June 29	98	39	0.85″	
June 30	98	40	2.06″	
July 1	100	41	1.47″	
July 2	99	42	2.23″	
July 3	98	41	1.73″	
July 4	101	41	1.16″	
July 5	**105**	41	1.69″	
July 6	99	39	2.10″	
July 7	100	41	1.69″	
July 8	100	40	1.01″	
July 9	97	42	1.73″	
July 10	102	41	1.13″	
July 11	97	42	1.20″	
July 12	101	N/A	1.24″	
July 13	**105**	41	2.10″	
July 14	102	43	2.18″	
July 15	98	42	1.39″	
July 16	99	43	2.90″	
July 17	98	44	4.05″	
July 18	99	37	2.70″	
July 19	100	42	1.32″	
July 20	100	37	1.26″	
July 21	103	41	2.95″	
July 22	100	41	2.10″	
July 23	102	41	2.03″	
July 24	104	40	2.19″	
July 25	102	46	1.58″	
July 26	98	44	1.30″	
July 27	100	43	1.69″	
July 28	100	39	1.50″	
July 29	103	41	3.00″	
July 30	100	42	0.97″	
July 31	101	45	1.85″	
August 1	98	N/A	0.66″	
August 2	100	38	3.35″	
August 3	100	41	2.53″	
August 4	98	42	2.67″	
August 5	100	42	1.70″	
August 6	104	43	1.35″	
August 7	100	40	2.82″	
August 8	95	33	2.97″	
August 9	94	41	2.37″	

Bloomingdale *(continued)*

	RECORD HIGH	RECORD LOW	RECORD PRECIPITATION	RECORD SNOWFALL
August 10	97	42	1.40″	
August 11	95	38	2.27″	
August 12	96	39	1.59″	
August 13	98	42	0.92″	
August 14	96	41	2.30″	
August 15	98	40	1.52″	
August 16	94	39	1.69″	
August 17	98	39	2.26″	
August 18	97	41	1.29″	
August 19	98	40	1.69″	
August 20	96	40	1.89″	
August 21	98	42	3.10″	
August 22	97	41	3.29″	
August 23	96	42	1.93″	
August 24	96	38	1.13″	
August 25	92	41	1.79″	
August 26	93	41	2.11″	
August 27	95	40	2.73″	
August 28	94	38	1.45″	
August 29	94	36	1.95″	
August 30	94	39	3.25″	
August 31	97	38	1.49″	
September 1	98	38	**9.78″**	
September 2	99	38	2.15″	
September 3	95	35	1.40″	
September 4	94	N/A	1.63″	
September 5	95	34	2.58″	
September 6	97	35	1.52″	
September 7	96	35	1.25″	
September 8	96	N/A	1.40″	
September 9	93	36	2.12″	
September 10	93	N/A	1.58″	
September 11	92	32	2.29″	
September 12	93	36	3.20″	
September 13	97	29	2.59″	
September 14	98	34	4.67″	
September 15	100	34	2.21″	
September 16	91	N/A	2.20″	
September 17	96	33	1.95″	
September 18	92	30	2.04″	
September 19	92	31	1.89″	
September 20	90	32	1.48″	
September 21	94	32	2.02″	
September 22	91	28	1.25″	
September 23	88	32	1.90″	
September 24	87	27	1.04″	
September 25	92	27	4.57″	
September 26	88	31	1.66″	
September 27	92	26	2.59″	
September 28	87	26	1.05″	
September 29	94	24	2.09″	

	RECORD HIGH	RECORD LOW	RECORD PRECIPITATION	RECORD SNOWFALL
September 30	86	29	1.52″	
October 1	88	27	1.14″	
October 2	89	27	1.03″	
October 3	88	24	1.75″	
October 4	86	23	1.47″	
October 5	86	24	1.47″	
October 6	86	24	2.35″	
October 7	89	27	1.44″	
October 8	88	27	0.85″	
October 9	88	22	1.12″	2.0″
October 10	84	23	1.72″	
October 11	85	20	2.38″	1.5″
October 12	88	26	2.02″	1.0″
October 13	85	24	0.92″	1.0″
October 14	85	23	1.68″	0.3″
October 15	83	23	1.22″	
October 16	88	26	1.25″	
October 17	84	23	2.22″	
October 18	84	23	1.75″	
October 19	83	16	1.75	
October 20	84	22	0.92″	6.0″
October 21	86	19	1.83″	
October 22	85	21	1.52″	
October 23	80	22	1.98″	0.3″
October 24	79	22	1.41″	5.0″
October 25	79	20	0.90″	6.2″
October 26	82	23	1.16″	4.2″
October 27	78	19	1.09″	5.0″
October 28	79	20	0.64″	6.0″
October 29	77	20	1.20″	2.0″
October 30	79	19	1.38″	13.8″
October 31	80	18	0.94″	3.3″
November 1	80	21	1.37″	4.0″
November 2	77	21	1.21″	3.0″
November 3	76	18	1.01″	3.0″
November 4	75	11	1.70″	8.0″
November 5	75	16	1.50″	4.0″
November 6	76	17	1.63″	3.0″
November 7	78	19	1.05″	5.0″
November 8	70	13	0.90″	4.0″
November 9	72	14	1.20″	4.5″
November 10	74	15	1.61″	8.0″
November 11	71	15	1.90″	10.0″
November 12	71	10	0.98″	3.5″
November 13	68	8	0.99″	3.5″
November 14	72	10	1.70″	8.5″
November 15	67	8	1.45″	11.0″
November 16	72	10	0.94″	4.0″
November 17	72	8	1.41″	4.0″
November 18	72	4	0.85″	7.0″

Bloomingdale *(continued)*

	RECORD HIGH	RECORD LOW	RECORD PRECIPITATION	RECORD SNOWFALL
November 19	74	8	1.51″	5.0″
November 20	71	13	1.51″	8.0″
November 21	69	2	0.99″	8.0″
November 22	67	8	0.75″	5.6″
November 23	70	7	0.80″	12.0″
November 24	68	2	1.00″	7.0″
November 25	65	–12	0.91″	9.0″
November 26	62	11	1.16″	5.0″
November 27	65	9	1.86″	5.8″
November 28	70	9	2.20″	6.5″
November 29	69	4	0.95″	5.5″
November 30	67	5	1.45″	14.0″
December 1	64	–16	0.82″	7.0″
December 2	67	–4	0.80″	9.0″
December 3	67	–7	2.67″	5.0″
December 4	68	–10	0.75″	9.0″
December 5	64	8	0.97″	9.0″
December 6	71	5	1.25″	12.5″
December 7	62	4	0.92″	10.0″
December 8	60	–2	0.73″	12.0″
December 9	60	–2	1.00″	12.0″
December 10	58	–1	1.23″	**20.0″**
December 11	62	–5	1.34″	17.0″
December 12	62	–4	1.80″	12.0″
December 13	60	–8	1.08″	12.0″
December 14	63	–7	0.80″	8.0″
December 15	60	–8	1.43″	8.0″
December 16	59	–3	1.50″	10.0″
December 17	63	–10	1.70″	8.0″
December 18	61	–12	0.63″	11.5″
December 19	57	–1	0.90″	9.0″
December 20	57	–5	1.00″	11.0″
December 21	58	–8	1.16″	8.0″
December 22	57	–12	0.93″	10.0″
December 23	56	–13	1.50″	8.0″
December 24	58	–12	0.82″	8.0″
December 25	59	–9	1.72″	9.0″
December 26	66	–14	0.75″	6.0″
December 27	54	–7	0.77″	8.0″
December 28	61	–7	1.99″	8.0″
December 29	66	–9	0.82″	15.0″
December 30	63	–16	1.34″	7.5″
December 31	60	–18	1.50″	6.0″

Coldwater (Climate Record Begins 1897)

	RECORD HIGH	RECORD LOW	RECORD PRECIPITATION	RECORD SNOWFALL
January 1	55	–9	1.54″	8.0″
January 2	56	–10	1.75″	10.0″
January 3	58	–15	0.89″	11.0″
January 4	58	**–23**	1.44″	4.0″
January 5	60	–11	0.97″	4.0″
January 6	57	–15	1.15″	6.0″
January 7	63	–11	1.08″	4.0″
January 8	61	–14	2.25″	3.0″
January 9	57	–7	0.49″	6.0″
January 10	56	–14	0.90″	5.5″
January 11	53	–16	0.50″	5.0″
January 12	63	–19	2.12″	5.0″
January 13	60	–14	1.03″	12.2″
January 14	58	–15	1.00″	5.4″
January 15	59	–18	0.62″	7.0″
January 16	58	–18	0.30″	7.0″
January 17	63	–18	0.53″	4.5″
January 18	59	–18	0.90″	5.0″
January 19	60	–17	1.13″	6.0″
January 20	69	–18	1.00″	4.0″
January 21	70	–19	1.10″	5.5″
January 22	59	–11	1.23″	6.0″
January 23	61	–18	0.64″	4.5″
January 24	61	–18	0.90″	4.0″
January 25	67	–19	0.65″	6.0″
January 26	65	–18	1.16″	**17.0″**
January 27	62	–10	0.89″	8.0″
January 28	55	–11	0.64″	6.0″
January 29	54	–10	1.12″	3.5″
January 30	53	–10	1.37″	10.0″
January 31	61	–13	1.21″	9.0″
February 1	55	–10	1.66″	5.0″
February 2	50	–14	0.50″	3.0″
February 3	59	–16	0.60″	6.0″
February 4	55	–12	1.28″	7.0″
February 5	53	–21	1.25″	7.0″
February 6	56	–8	0.85″	5.0″
February 7	54	–15	1.09″	5.5″
February 8	62	–15	0.80″	8.0″
February 9	57	–17	1.09″	5.0″
February 10	59	–17	0.96″	6.0″
February 11	62	–11	1.30″	4.0″
February 12	67	–17	1.20″	3.5″
February 13	62	–15	0.55″	6.0″
February 14	62	–17	1.25″	4.5″

Coldwater *(continued)*

	RECORD HIGH	RECORD LOW	RECORD PRECIPITATION	RECORD SNOWFALL
February 15	65	–17	0.98″	3.0″
February 16	63	–11	2.52″	5.4″
February 17	57	–15	0.71″	4.0″
February 18	59	–15	0.67″	5.0″
February 19	63	–11	1.85″	4.0″
February 20	64	–22	0.67″	6.0″
February 21	67	–10	0.93″	6.0″
February 22	67	–11	2.18″	5.0″
February 23	65	–7	1.60″	3.2″
February 24	61	–10	1.05″	7.0″
February 25	66	–11	1.36″	7.0″
February 26	66	–20	0.97″	4.5″
February 27	66	–10	1.06″	5.0″
February 28	57	–9	1.10″	11.0″
February 29	58	–2	0.45″	3.0″
March 1	59	–13	1.00″	4.0″
March 2	64	–6	0.56″	4.0″
March 3	69	–5	0.95″	5.0″
March 4	65	–1	1.30″	6.0″
March 5	69	–3	1.20″	12.0″
March 6	68	–1	1.40″	3.0″
March 7	73	–1	1.41″	4.0″
March 8	75	–10	1.85″	5.5″
March 9	76	–6	1.33″	12.0″
March 10	68	1	1.25″	5.5″
March 11	71	3	0.66″	3.0″
March 12	73	–5	1.70″	4.0″
March 13	76	4	1.32″	3.0″
March 14	74	3	0.82″	7.0″
March 15	77	5	1.55″	6.0″
March 16	75	–1	0.96″	5.0″
March 17	71	–3	0.90″	6.0″
March 18	76	0	0.72″	3.0″
March 19	76	1	1.66″	6.0″
March 20	75	8	0.75″	5.0″
March 21	75	9	0.89″	9.0″
March 22	80	7	1.00″	6.2″
March 23	76	3	0.64″	4.0″
March 24	83	–3	1.65″	0.3″
March 25	77	–1	1.64″	8.0″
March 26	78	6	1.31″	8.0″
March 27	79	–4	1.40″	6.0″
March 28	81	11	1.08″	4.0″
March 29	80	10	1.15″	3.6″
March 30	76	11	1.08″	4.0″
March 31	78	5	1.83″	1.5″
April 1	77	8	1.78″	6.0″
April 2	80	18	1.08″	5.5″
April 3	77	14	1.80″	5.0″
April 4	77	15	1.21″	4.0″

	RECORD HIGH	RECORD LOW	RECORD PRECIPITATION	RECORD SNOWFALL
April 5	80	14	2.70″	7.0″
April 6	80	14	1.82″	10.0″
April 7	80	6	1.50″	4.0″
April 8	76	6	1.01″	1.5″
April 9	80	15	1.06″	1.2″
April 10	82	16	1.16″	2.0″
April 11	86	21	1.53″	4.0″
April 12	85	14	1.30″	1.0″
April 13	86	18	0.75″	1.0″
April 14	80	19	1.02″	1.6″
April 15	83	16	1.70″	1.0″
April 16	84	20	1.00″	3.0″
April 17	82	22	1.55″	2.0″
April 18	85	20	1.85″	2.0″
April 19	84	21	1.10″	
April 20	82	15	2.20″	1.0″
April 21	84	21	1.78″	4.0″
April 22	86	25	1.28″	
April 23	86	20	1.19″	4.0″
April 24	86	24	3.07″	1.0″
April 25	85	22	1.82″	1.5″
April 26	89	25	1.11″	4.0″
April 27	84	27	0.88″	1.0″
April 28	83	25	1.51″	
April 29	88	27	2.28″	
April 30	88	26	1.86″	
May 1	86	24	1.60″	1.0″
May 2	87	28	1.35″	2.5″
May 3	86	28	1.16″	
May 4	86	23	0.90″	1.5″
May 5	87	25	1.03″	
May 6	89	20	2.63″	
May 7	86	25	1.20″	
May 8	86	27	1.19″	
May 9	88	26	1.90″	4.0″
May 10	89	24	1.46″	0.5″
May 11	86	25	2.54″	1.0″
May 12	86	28	2.22″	
May 13	86	28	1.56″	
May 14	87	28	2.50″	
May 15	88	30	3.15″	
May 16	89	29	2.21″	
May 17	90	29	1.77″	
May 18	91	28	1.99″	
May 19	93	29	1.76″	
May 20	90	29	1.15″	
May 21	90	27	1.47″	
May 22	90	30	1.42″	
May 23	91	28	2.06″	
May 24	88	30	1.81″	
May 25	91	29	1.58″	

Coldwater *(continued)*

	RECORD HIGH	RECORD LOW	RECORD PRECIPITATION	RECORD SNOWFALL
May 26	93	32	1.62″	
May 27	96	31	2.25″	
May 28	96	31	1.24″	
May 29	93	34	1.40″	
May 30	92	33	4.48″	
May 31	92	34	3.45″	
June 1	102	34	2.16″	
June 2	98	35	2.01″	
June 3	96	34	2.20″	
June 4	98	38	2.00″	
June 5	97	37	1.24″	
June 6	96	36	2.06″	
June 7	99	37	2.12″	
June 8	98	35	1.56″	
June 9	97	35	1.14″	
June 10	98	35	1.34″	
June 11	98	35	1.15″	
June 12	95	39	2.10″	
June 13	93	40	1.84″	
June 14	96	37	1.20″	
June 15	94	40	1.42″	
June 16	96	38	1.57″	
June 17	95	42	1.22″	
June 18	95	42	2.50″	
June 19	97	42	1.80″	
June 20	99	40	1.80″	
June 21	99	38	2.88″	
June 22	99	38	2.35″	
June 23	95	38	2.60″	
June 24	97	40	1.65″	
June 25	102	37	4.55″	
June 26	94	42	**5.37″**	
June 27	99	40	2.35″	
June 28	102	44	1.18″	
June 29	98	42	1.60″	
June 30	99	40	1.56″	
July 1	98	40	1.90″	
July 2	98	41	1.75″	
July 3	100	43	2.80″	
July 4	100	40	3.40″	
July 5	101	44	1.18″	
July 6	99	42	0.98″	
July 7	100	45	2.32″	
July 8	102	42	2.48″	
July 9	100	42	2.24″	
July 10	102	42	1.11″	
July 11	101	42	1.02″	
July 12	103	43	1.55″	
July 13	105	45	1.47″	
July 14	105	44	2.00″	
July 15	98	43	2.25″	
July 16	98	42	2.50″	
July 17	98	41	1.62″	
July 18	99	47	2.40″	
July 19	103	46	2.00″	
July 20	102	40	2.50″	
July 21	106	43	2.00″	
July 22	104	45	2.80″	
July 23	106	45	2.00″	
July 24	**108**	45	2.21″	
July 25	104	43	2.20″	
July 26	99	45	1.75″	
July 27	100	43	1.65″	
July 28	99	48	3.25″	
July 29	101	46	4.75″	
July 30	102	43	2.12″	
July 31	97	44	2.00″	
August 1	101	45	2.95″	
August 2	99	41	3.60″	
August 3	98	41	3.20″	
August 4	96	44	1.18″	
August 5	100	44	2.03″	
August 6	103	43	1.85″	
August 7	99	46	2.88″	
August 8	96	43	1.48″	
August 9	99	39	0.92″	
August 10	96	41	2.20″	
August 11	97	42	1.66″	
August 12	95	42	2.40″	
August 13	99	39	1.35″	
August 14	97	39	1.77″	
August 15	95	38	1.63″	
August 16	97	36	1.70″	
August 17	96	41	2.10″	
August 18	97	38	1.38″	
August 19	97	38	3.50″	
August 20	99	41	1.32″	
August 21	99	39	3.37″	
August 22	100	40	2.35″	
August 23	98	38	1.40″	
August 24	95	40	1.53″	
August 25	96	40	2.79″	
August 26	97	41	2.42″	
August 27	99	43	3.30″	
August 28	95	38	1.00″	
August 29	96	37	2.41″	
August 30	96	36	1.08″	
August 31	97	38	2.06″	
September 1	98	38	2.08″	
September 2	99	37	1.46″	

Coldwater *(continued)*

	RECORD HIGH	RECORD LOW	RECORD PRECIPITATION	RECORD SNOWFALL
September 3	97	33	1.75″	
September 4	96	38	2.50″	
September 5	95	35	1.38″	
September 6	94	34	2.24″	
September 7	98	30	0.86″	
September 8	99	36	1.56″	
September 9	96	38	3.00″	
September 10	98	36	1.50″	
September 11	95	32	1.10″	
September 12	95	35	3.18″	
September 13	95	33	3.77″	
September 14	95	33	3.28″	
September 15	97	36	1.25″	
September 16	95	34	1.30″	
September 17	91	29	3.08″	
September 18	90	32	1.25″	
September 19	92	30	3.90″	
September 20	89	32	0.75″	
September 21	92	28	2.70″	
September 22	90	32	1.21″	
September 23	90	29	1.10″	
September 24	91	29	1.65″	
September 25	91	32	1.04″	
September 26	90	28	1.90″	
September 27	88	29	1.36″	
September 28	90	28	1.20″	
September 29	94	27	2.00″	3.0″
September 30	88	28	1.15″	
October 1	89	22	1.30″	
October 2	88	25	1.30″	
October 3	89	23	2.75″	
October 4	89	29	1.31″	
October 5	88	26	1.50″	
October 6	89	24	1.87″	
October 7	86	22	1.70″	
October 8	87	28	0.70″	
October 9	85	24	1.60″	
October 10	88	23	1.52″	
October 11	84	19	1.64″	
October 12	88	23	5.30″	0.2″
October 13	88	20	1.30″	
October 14	83	27	1.83″	
October 15	86	24	2.27″	
October 16	84	25	2.01″	
October 17	83	22	1.47″	
October 18	83	17	1.96″	0.1″
October 19	84	19	1.50″	4.0″
October 20	85	20	0.69″	2.5″
October 21	85	16	1.08″	2.0″
October 22	84	22	1.30″	
October 23	83	21	2.35″	5.0″

	RECORD HIGH	RECORD LOW	RECORD PRECIPITATION	RECORD SNOWFALL
October 24	80	19	1.20″	2.0″
October 25	80	20	1.50″	1.0″
October 26	80	22	1.00″	0.5″
October 27	78	16	1.40″	5.0″
October 28	80	19	0.83″	2.0″
October 29	80	18	1.37″	
October 30	78	18	0.90″	1.5″
October 31	80	17	1.08″	1.0″
November 1	78	18	1.36″	0.5″
November 2	78	18	1.02″	4.0″
November 3	75	14	0.95″	6.0″
November 4	74	15	1.40″	2.5″
November 5	73	11	1.36″	2.0″
November 6	75	16	1.23″	4.0″
November 7	74	14	1.98″	3.5″
November 8	76	14	1.35″	2.3″
November 9	72	12	1.36″	4.0″
November 10	68	16	1.40″	2.5″
November 11	75	15	1.10″	4.5″
November 12	70	14	1.28″	1.0″
November 13	72	9	1.57″	4.5″
November 14	69	10	0.96″	2.0″
November 15	69	10	1.09″	4.0″
November 16	71	5	1.31″	5.5″
November 17	70	10	1.20″	3.0″
November 18	69	5	1.25″	4.5″
November 19	75	5	2.00″	6.0″
November 20	78	10	1.50″	3.0″
November 21	69	12	1.30″	3.5″
November 22	66	10	2.00″	1.8″
November 23	68	6	1.36″	3.5″
November 24	63	−4	0.73″	4.0″
November 25	64	−3	1.12″	4.0″
November 26	64	0	1.19″	6.0″
November 27	67	1	1.04″	4.0″
November 28	70	3	1.40″	3.0″
November 29	70	3	1.70″	7.0″
November 30	62	−3	0.98″	2.0″
December 1	66	6	1.01″	4.0″
December 2	67	−14	0.66″	5.0″
December 3	65	−14	1.08″	3.0″
December 4	63	−1	1.27″	6.0″
December 5	67	−2	1.29″	5.0″
December 6	62	−3	1.47″	6.0″
December 7	65	−4	1.33″	8.0″
December 8	63	−13	1.60″	7.0″
December 9	65	−9	0.90″	4.0″
December 10	65	−4	0.90″	4.5″
December 11	59	−10	1.10″	6.5″
December 12	62	−5	1.24″	6.0″

Coldwater *(continued)*

	RECORD HIGH	RECORD LOW	RECORD PRECIPITATION	RECORD SNOWFALL
December 13	62	–7	1.55″	5.0″
December 14	65	–16	0.70″	7.0″
December 15	62	–17	1.46″	3.8″
December 16	62	–8	0.78″	7.0″
December 17	54	–5	1.56″	4.5″
December 18	55	–5	1.00″	4.0″
December 19	56	–8	0.83″	12.0″
December 20	57	–7	0.65″	3.0″
December 21	61	–12	1.98″	6.0″
December 22	57	–15	0.90″	4.0″
December 23	56	–13	2.00″	8.0″
December 24	59	–10	2.64″	3.5″
December 25	67	–14	0.70″	6.0″
December 26	52	–11	0.80″	6.0″
December 27	60	–8	1.04″	5.0″
December 28	63	–10	0.60″	6.0″
December 29	62	–7	2.22″	2.8″
December 30	58	–8	1.54″	3.0″
December 31	60	–14	1.02″	6.0″

Detroit (Climate Record Begins 1870)

	RECORD HIGH	RECORD LOW	RECORD PRECIPITATION	RECORD SNOWFALL
January 1	65	–7	1.41″	7.0″
January 2	62	–12	1.00″	10.6″
January 3	59	–15	1.11″	5.8″
January 4	61	–12	1.59″	2.8″
January 5	59	–10	0.91″	4.6″
January 6	62	–7	0.74″	10.1″
January 7	57	–5	0.62″	3.1″
January 8	62	–10	1.31″	2.8″
January 9	55	–12	0.92″	6.2″
January 10	59	–15	0.79″	6.4″
January 11	66	–9	0.96″	5.2″
January 12	63	–16	1.76″	8.0″
January 13	62	–9	0.98″	8.5″
January 14	58	–9	0.78″	11.1″
January 15	59	–12	0.87″	4.0″
January 16	55	–15	1.61″	4.2″
January 17	63	–15	0.68″	4.6″
January 18	58	–18	0.87″	5.2″
January 19	60	–20	0.94″	5.0″
January 20	64	–14	0.92″	2.7″
January 21	65	–21	0.88″	6.0″
January 22	62	–10	1.19″	12.2″
January 23	61	–12	0.69″	3.2″
January 24	61	–13	0.49″	3.6″
January 25	67	–16	0.55″	4.0″
January 26	64	–9	1.39″	7.8″
January 27	61	–6	0.77″	4.3″
January 28	57	–8	0.84″	5.0″
January 29	58	–12	0.84″	4.5″
January 30	53	–4	1.15″	7.3″
January 31	60	–7	1.48″	10.0″
February 1	54	–7	1.11″	9.8″
February 2	52	–12	0.49″	4.2″
February 3	54	–15	1.12″	7.2″
February 4	63	–10	0.98″	10.0″
February 5	51	–16	1.20″	5.0″
February 6	59	–7	1.23″	12.3″
February 7	53	–10	1.04″	5.6″
February 8	62	–11	0.74″	7.4″
February 9	56	–20	2.41″	5.0″
February 10	59	–13	1.24″	8.1″
February 11	70	–10	0.82″	7.3″
February 12	62	–14	1.17″	10.5″
February 13	63	–12	1.15″	8.0″
February 14	57	–15	1.92″	5.4″
February 15	66	–10	0.96″	5.0″
February 16	60	–9	1.55″	7.7″
February 17	59	–7	2.24″	3.3″
February 18	62	–9	1.25″	8.0″
February 19	64	–4	2.26″	12.6″
February 20	61	–8	1.07″	5.6″
February 21	63	–2	0.87″	9.0″
February 22	65	–5	2.28″	5.8″
February 23	64	–8	1.47″	2.5″
February 24	60	–11	0.63″	5.1″
February 25	66	–2	0.91″	9.3″
February 26	69	–1	0.75″	7.1″
February 27	63	–5	0.99″	5.2″
February 28	60	–1	1.24″	12.4″
February 29	63	–6	1.16″	1.2″
March 1	64	1	1.29″	6.2″
March 2	63	–4	0.82″	2.7″
March 3	67	–4	1.71″	1.5″
March 4	69	1	1.18″	11.8″
March 5	68	–7	0.98″	9.8″
March 6	67	2	0.73″	4.0″
March 7	76	2	1.06″	4.0″
March 8	80	–1	0.95″	7.6″

Detroit *(continued)*

	RECORD HIGH	RECORD LOW	RECORD PRECIPITATION	RECORD SNOWFALL
March 9	68	1	1.22″	7.7″
March 10	70	4	0.97″	7.5″
March 11	72	7	0.92″	5.0″
March 12	75	4	1.13″	2.4″
March 13	73	5	1.58″	5.0″
March 14	77	2	1.54″	8.5″
March 15	77	5	1.05″	8.6″
March 16	74	4	1.15″	3.7″
March 17	75	–2	1.03″	8.4″
March 18	72	7	0.99″	6.8″
March 19	76	4	0.93″	7.5″
March 20	73	–2	0.88″	5.6″
March 21	73	–1	1.03″	6.0″
March 22	81	3	0.93″	9.5″
March 23	73	2	0.91″	5.3″
March 24	81	3	0.96″	1.6″
March 25	78	4	1.73″	6.0″
March 26	75	11	1.08″	3.6″
March 27	79	3	1.35″	6.3″
March 28	82	4	1.69″	2.5″
March 29	78	4	0.79″	4.6″
March 30	81	10	1.76″	7.1″
March 31	80	6	1.52″	4.7″
April 1	79	14	1.30″	3.0″
April 2	83	17	1.44″	2.0″
April 3	77	14	1.06″	4.2″
April 4	74	9	1.06″	2.7″
April 5	79	16	2.59″	5.0″
April 6	83	18	2.41″	**24.5**″
April 7	83	10	0.86″	6.0″
April 8	79	11	1.72″	0.8″
April 9	78	19	1.40″	2.8″
April 10	86	20	1.21″	3.0″
April 11	87	22	1.60″	2.5″
April 12	89	18	1.24″	1.8″
April 13	87	19	1.52″	2.5″
April 14	81	20	1.14″	1.2″
April 15	85	21	1.48″	2.5″
April 16	86	17	1.19″	3.3″
April 17	85	10	1.57″	4.5″
April 18	86	8	0.93″	0.3″
April 19	83	23	1.38″	3.0″
April 20	82	20	3.58″	1.2″
April 21	86	21	1.13″	0.3″
April 22	87	20	1.23″	1.0″
April 23	88	23	0.96″	1.2″
April 24	86	27	1.57″	3.1″
April 25	87	26	1.41″	0.4″
April 26	86	27	1.10″	0.1″
April 27	84	28	1.40″	
April 28	84	27	1.93″	

	RECORD HIGH	RECORD LOW	RECORD PRECIPITATION	RECORD SNOWFALL
April 29	83	26	1.99″	3.0″
April 30	87	28	1.55″	
May 1	87	26	1.46″	0.3″
May 2	85	29	1.66″	0.1″
May 3	88	28	1.01″	0.1″
May 4	89	28	1.05″	0.2″
May 5	90	29	0.75″	
May 6	90	31	1.75″	
May 7	86	27	0.97″	
May 8	86	30	0.83″	
May 9	90	29	2.12″	6.0″
May 10	90	25	2.33″	0.5″
May 11	87	30	1.91″	
May 12	91	32	1.48″	
May 13	89	30	1.48″	1.5″
May 14	91	34	1.39″	
May 15	92	33	1.46″	
May 16	92	32	1.09″	
May 17	93	31	1.22″	
May 18	93	32	1.50″	
May 19	91	32	1.68″	
May 20	91	33	1.56″	
May 21	92	33	1.69″	2.3″
May 22	90	33	2.06″	2.7″
May 23	89	34	1.69″	
May 24	87	33	1.26″	
May 25	89	35	1.52″	
May 26	92	36	2.56″	
May 27	91	35	1.68″	
May 28	93	35	2.02″	
May 29	92	32	1.65″	
May 30	93	36	2.27″	
May 31	95	34	1.98″	
June 1	97	36	2.02″	
June 2	95	38	2.41″	
June 3	96	39	1.63″	
June 4	95	40	1.94″	
June 5	96	38	2.00″	
June 6	94	41	3.07″	
June 7	93	38	2.55″	
June 8	100	39	2.00″	
June 9	93	40	2.03″	
June 10	95	38	1.65″	
June 11	97	36	1.71″	
June 12	94	42	0.68″	
June 13	96	42	1.80″	
June 14	96	42	2.00″	
June 15	95	46	1.52″	
June 16	97	43	2.21″	
June 17	99	43	1.62″	

Detroit *(continued)*

	RECORD HIGH	RECORD LOW	RECORD PRECIPITATION	RECORD SNOWFALL
June 18	99	47	1.52″	
June 19	95	46	0.67″	
June 20	97	43	1.96″	
June 21	96	42	2.23″	
June 22	98	44	2.15″	
June 23	95	42	1.05″	
June 24	97	45	2.54″	
June 25	104	43	2.17″	
June 26	100	44	2.10″	
June 27	99	48	1.79″	
June 28	104	49	1.56″	
June 29	96	48	1.32″	
June 30	96	47	1.08″	
July 1	98	47	1.52″	
July 2	98	46	1.09″	
July 3	100	48	2.17″	
July 4	98	49	2.38″	
July 5	96	47	1.09″	
July 6	100	42	1.12″	
July 7	101	44	4.34″	
July 8	104	45	2.80″	
July 9	102	50	2.51″	
July 10	102	51	2.05″	
July 11	101	47	3.08″	
July 12	100	43	3.19″	
July 13	102	48	1.48″	
July 14	104	46	1.26″	
July 15	102	50	1.94″	
July 16	102	49	0.93″	
July 17	101	46	1.20″	
July 18	99	49	2.70″	
July 19	100	48	2.61″	
July 20	97	48	2.04″	
July 21	97	50	1.19″	
July 22	96	49	1.73″	
July 23	98	50	1.31″	
July 24	**105**	52	2.34″	
July 25	99	51	1.16″	
July 26	97	50	2.17″	
July 27	100	48	2.62″	
July 28	100	48	3.54″	
July 29	99	50	2.52″	
July 30	98	50	1.75″	
July 31	96	48	**4.74″**	
August 1	97	48	1.17″	
August 2	99	48	1.46″	
August 3	96	46	1.83″	
August 4	98	47	3.90″	
August 5	96	45	2.33″	
August 6	104	49	2.51″	
August 7	100	47	1.26″	
August 8	99	47	2.03″	
August 9	96	46	1.03″	
August 10	99	45	1.30″	
August 11	99	47	2.06″	
August 12	99	46	2.16″	
August 13	96	47	2.18″	
August 14	99	48	1.86″	
August 15	97	46	1.31″	
August 16	95	43	1.57″	
August 17	100	46	4.51″	
August 18	95	46	1.25″	
August 19	95	44	1.90″	
August 20	96	48	1.50″	
August 21	100	46	2.72″	
August 22	101	45	1.99″	
August 23	96	45	2.29″	
August 24	95	43	1.05″	
August 25	98	48	1.76″	
August 26	96	47	2.11″	
August 27	98	47	2.27″	
August 28	97	43	1.77″	
August 29	96	38	2.38″	
August 30	97	41	2.24″	
August 31	97	46	2.02″	
September 1	98	42	2.02″	
September 2	100	44	1.16″	
September 3	100	43	3.21″	
September 4	92	40	2.10″	
September 5	99	42	0.97″	
September 6	95	38	2.73″	
September 7	97	43	1.40″	
September 8	98	39	0.91″	
September 9	94	37	1.82″	
September 10	94	40	1.65″	
September 11	95	39	3.71″	
September 12	96	39	1.41″	
September 13	97	40	2.97″	
September 14	98	37	1.58″	
September 15	100	40	1.65″	
September 16	98	38	1.38″	
September 17	93	36	1.20″	
September 18	92	37	1.06″	
September 19	93	38	1.64″	
September 20	92	36	1.35″	
September 21	92	35	1.73″	
September 22	91	30	1.84″	
September 23	89	29	1.28″	
September 24	89	33	1.73″	
September 25	93	34	1.40″	
September 26	91	30	2.08″	

Detroit *(continued)*

	RECORD HIGH	RECORD LOW	RECORD PRECIPITATION	RECORD SNOWFALL
September 27	88	34	1.28″	
September 28	87	33	2.13″	
September 29	89	32	1.53″	
September 30	85	30	1.30″	
October 1	88	31	1.55″	
October 2	86	29	2.00″	
October 3	89	24	3.29″	
October 4	89	32	1.06″	
October 5	88	31	2.10″	
October 6	91	30	2.20″	
October 7	92	28	1.50″	
October 8	89	25	1.02″	
October 9	86	28	1.43″	
October 10	84	29	1.03″	
October 11	86	25	3.27″	
October 12	84	26	1.25″	0.2″
October 13	83	26	1.57″	0.4″
October 14	83	27	1.78″	
October 15	86	24	1.03″	
October 16	85	26	1.92″	0.1″
October 17	85	23	1.70″	0.9″
October 18	84	24	1.54″	0.1″
October 19	85	22	2.02″	2.7″
October 20	80	19	1.09″	0.4″
October 21	81	17	0.96″	
October 22	81	25	1.98″	
October 23	83	22	2.08″	0.1″
October 24	81	22	1.07″	
October 25	82	24	1.09″	
October 26	83	22	1.29″	0.3″
October 27	77	22	1.48″	2.3″
October 28	78	21	0.73″	2.0″
October 29	77	22	0.74″	
October 30	76	20	1.29″	0.3″
October 31	79	21	0.87″	0.1″
November 1	81	21	1.59″	
November 2	75	22	1.02″	4.7″
November 3	75	16	0.90″	1.3″
November 4	75	17	1.41″	0.1″
November 5	74	17	1.37″	1.6″
November 6	75	18	1.64″	5.6″
November 7	70	20	1.00″	1.6″
November 8	71	14	1.57″	0.9″
November 9	75	18	1.38″	5.3″
November 10	68	19	1.57″	3.4″
November 11	71	19	1.16″	4.1″
November 12	69	16	1.59″	3.1″
November 13	70	12	1.02″	3.7″
November 14	70	11	1.17″	3.0″
November 15	71	10	2.30″	6.2″
November 16	69	8	1.50″	7.2″
November 17	70	13	1.46″	4.4″
November 18	69	11	0.97″	2.0″
November 19	68	9	1.07″	3.2″
November 20	70	12	1.02″	2.6″
November 21	67	3	0.81″	1.4″
November 22	69	0	2.59″	1.1″
November 23	69	8	1.24″	2.3″
November 24	62	7	0.82″	2.6″
November 25	63	10	1.52″	3.8″
November 26	65	10	1.23″	2.1″
November 27	69	9	1.69″	5.6″
November 28	68	8	1.52″	2.0″
November 29	67	5	1.43″	3.7″
November 30	65	4	1.05″	3.8″
December 1	64	8	1.69″	18.4″
December 2	67	−2	1.35″	5.8″
December 3	68	−3	1.58″	9.0″
December 4	64	3	0.90″	7.0″
December 5	67	0	1.42″	7.0″
December 6	69	7	1.12″	6.6″
December 7	62	1	1.47″	6.6″
December 8	66	0	1.32″	5.8″
December 9	58	−5	1.09″	6.0″
December 10	64	−9	0.84″	5.2″
December 11	62	−9	0.92″	5.5″
December 12	61	−2	1.39″	4.0″
December 13	60	1	1.15″	5.8″
December 14	65	−1	1.35″	10.0″
December 15	61	−1	1.45″	5.7″
December 16	65	−3	0.85″	2.8″
December 17	56	−7	0.85″	2.4″
December 18	54	−3	0.94″	8.0″
December 19	58	−6	1.08″	8.7″
December 20	57	−10	1.01″	4.4″
December 21	61	−14	2.17″	5.0″
December 22	56	**−24**	1.09″	2.6″
December 23	56	−9	1.30″	8.4″
December 24	61	−14	1.95″	6.5″
December 25	64	−10	1.16″	7.9″
December 26	55	−8	1.06″	8.6″
December 27	58	−4	1.14″	5.2″
December 28	64	−4	0.99″	7.2″
December 29	65	−11	1.34″	3.6″
December 30	59	−10	1.31″	2.5″
December 31	65	−5	0.98″	4.5″

East Tawas (Climate Record Begins 1890)

	RECORD HIGH	RECORD LOW	RECORD PRECIPITATION	RECORD SNOWFALL
January 1	58	–20	1.30″	11.7″
January 2	45	–5	0.69″	6.9″
January 3	49	–10	1.41″	13.6″
January 4	52	–20	2.10″	12.0″
January 5	49	–10	0.60″	5.0″
January 6	52	–12	0.90″	5.0″
January 7	47	–17	0.65″	6.0″
January 8	55	–12	0.82″	3.7″
January 9	49	–14	0.80″	6.0″
January 10	52	–17	1.20″	12.0″
January 11	47	–15	1.09″	4.0″
January 12	41	–12	1.09″	12.5″
January 13	52	–14	0.70″	7.0″
January 14	46	–12	1.25″	14.0″
January 15	45	–15	0.61″	6.2″
January 16	49	–18	0.41″	5.2″
January 17	49	–16	0.80″	4.0″
January 18	48	–19	0.63″	5.6″
January 19	46	–21	0.91″	6.0″
January 20	51	–17	1.85″	4.0″
January 21	51	–13	1.50″	6.4″
January 22	53	–15	0.82″	8.0″
January 23	50	–18	0.99″	5.8″
January 24	52	–17	1.02″	4.0″
January 25	55	–17	0.85″	5.0″
January 26	51	–17	1.60″	17.0″
January 27	52	–19	1.05″	10.3″
January 28	52	–15	0.67″	6.7″
January 29	48	–12	0.80″	8.0″
January 30	44	–25	1.04″	10.0″
January 31	48	–21	0.35″	5.7″
February 1	55	**–29**	0.93″	7.0″
February 2	49	–23	0.80″	8.0″
February 3	48	–20	0.67″	5.2″
February 4	55	–19	0.45″	6.0″
February 5	48	–27	1.50″	6.3″
February 6	48	–17	1.00″	8.0″
February 7	48	–21	0.79″	8.1″
February 8	52	–21	1.00″	8.1″
February 9	50	–22	1.20″	12.0″
February 10	48	–24	1.49″	6.0″
February 11	57	–21	1.31″	6.0″
February 12	58	–21	1.01″	7.1″
February 13	45	–15	0.82″	4.8″
February 14	48	–13	2.00″	**20.0**″
February 15	47	–18	1.21″	5.5″
February 16	61	–20	0.58″	4.0″
February 17	49	–26	1.51″	3.0″
February 18	49	–24	0.93″	4.0″
February 19	54	–19	0.60″	5.1″
February 20	52	**–29**	0.40″	4.0″
February 21	51	–20	1.03″	12.0″
February 22	51	–24	1.40″	8.6″
February 23	53	–20	1.30″	9.5″
February 24	54	–15	0.65″	5.0″
February 25	49	–16	1.31″	7.5″
February 26	53	–15	1.00″	10.0″
February 27	59	–13	1.37″	7.0″
February 28	49	–11	1.20″	4.0″
February 29	51	–5	0.23″	2.0″
March 1	53	–13	0.71″	4.8″
March 2	50	–19	1.54″	6.3″
March 3	54	–16	0.84″	5.0″
March 4	65	–11	0.93″	8.4″
March 5	59	–3	1.30″	6.5″
March 6	52	–15	0.85″	11.6″
March 7	57	–11	0.74″	5.0″
March 8	75	–18	0.54″	6.0″
March 9	67	–11	1.52″	9.0″
March 10	62	–7	0.95″	6.8″
March 11	61	–3	0.81″	4.0″
March 12	60	–8	1.29″	3.1″
March 13	70	–2	1.20″	7.0″
March 14	69	–3	1.20″	10.1″
March 15	69	–3	0.80″	6.0″
March 16	70	–2	0.79″	3.0″
March 17	74	–4	1.24″	6.0″
March 18	61	–2	1.13″	6.8″
March 19	70	–3	1.10″	10.0″
March 20	72	–11	0.88″	2.0″
March 21	68	–12	0.76″	6.0″
March 22	70	–1	1.24″	7.6″
March 23	62	–5	0.92″	3.0″
March 24	77	–1	0.62″	7.5″
March 25	70	–3	1.00″	5.2″
March 26	70	–2	1.36″	11.0″
March 27	78	6	2.00″	4.0″
March 28	78	–1	0.93″	3.0″
March 29	77	–6	0.76″	3.0″
March 30	72	8	0.91″	8.0″
March 31	75	–5	1.08″	4.0″
April 1	73	–7	1.40″	4.0″
April 2	80	7	0.90″	9.0″
April 3	74	6	1.18″	3.2″
April 4	71	0	1.80″	7.0″
April 5	69	10	0.93″	8.0″
April 6	79	6	1.40″	3.0″
April 7	81	6	1.61″	2.0″
April 8	84	11	0.61″	5.8″

East Tawas *(continued)*

	RECORD HIGH	RECORD LOW	RECORD PRECIPITATION	RECORD SNOWFALL
April 9	76	7	3.52″	1.5″
April 10	74	10	0.70″	5.0″
April 11	70	12	1.31″	2.0″
April 12	82	17	1.50″	2.9″
April 13	84	11	1.80″	2.7″
April 14	79	16	1.03″	6.0″
April 15	80	14	1.42″	4.1″
April 16	86	12	1.18″	5.5″
April 17	89	19	1.39″	
April 18	83	19	1.20″	1.0″
April 19	80	15	2.30″	0.1″
April 20	84	16	1.27″	1.5″
April 21	84	16	1.20″	2.0″
April 22	89	18	1.51″	1.5″
April 23	85	20	0.69″	2.0″
April 24	82	21	1.00″	1.1″
April 25	79	21	1.61″	
April 26	91	18	0.85″	3.0″
April 27	85	18	0.87″	
April 28	81	20	1.60″	1.0″
April 29	82	21	2.55″	
April 30	81	19	1.25″	0.9″
May 1	79	20	1.67″	0.3″
May 2	84	18	1.12″	1.0″
May 3	87	22	1.04″	3.0″
May 4	90	23	2.42″	
May 5	93	24	0.78″	
May 6	90	21	0.93″	1.0″
May 7	87	20	1.75″	
May 8	87	25	1.62″	
May 9	89	24	2.12″	4.0″
May 10	85	23	1.35″	
May 11	84	23	1.40″	
May 12	85	24	1.17″	
May 13	86	24	1.92″	
May 14	88	27	0.77″	2.0″
May 15	90	23	1.24″	
May 16	88	29	1.00″	
May 17	88	27	1.15″	
May 18	89	25	1.25″	
May 19	92	25	1.85″	
May 20	88	25	2.47″	
May 21	88	25	1.12″	
May 22	93	25	2.80″	
May 23	89	24	1.51″	
May 24	85	25	1.57″	
May 25	87	25	2.06″	
May 26	92	22	1.00″	
May 27	90	26	1.94″	
May 28	95	28	2.08″	
May 29	91	31	0.80″	

	RECORD HIGH	RECORD LOW	RECORD PRECIPITATION	RECORD SNOWFALL
May 30	90	28	1.51″	
May 31	92	31	1.28″	
June 1	93	28	1.75″	
June 2	94	31	1.81″	
June 3	88	31	2.40″	
June 4	88	31	1.06″	
June 5	97	31	1.29″	
June 6	99	31	1.53″	
June 7	95	34	1.30″	
June 8	92	28	1.65″	
June 9	92	30	1.41″	
June 10	90	32	1.24″	
June 11	91	28	1.99″	
June 12	98	35	1.16″	
June 13	95	38	1.18″	
June 14	92	31	3.10″	
June 15	98	37	1.35″	
June 16	90	33	1.00″	
June 17	93	35	1.26″	
June 18	92	34	1.70″	
June 19	96	36	1.06″	
June 20	101	36	1.81″	
June 21	92	33	0.98″	
June 22	92	35	0.72″	
June 23	91	38	2.02″	
June 24	97	32	2.06″	
June 25	96	34	2.70″	
June 26	99	38	3.64″	
June 27	95	38	1.24″	
June 28	97	40	1.29″	
June 29	100	35	0.97″	
June 30	101	38	3.33″	
July 1	97	39	2.35″	
July 2	100	38	0.62″	
July 3	97	39	1.60″	
July 4	95	33	1.83″	
July 5	97	35	2.02″	
July 6	99	37	0.98″	
July 7	98	42	1.01″	
July 8	**106**	38	3.48″	
July 9	**106**	38	1.69″	
July 10	102	42	2.55″	
July 11	96	38	1.08″	
July 12	94	42	0.90″	
July 13	96	37	1.06″	
July 14	96	41	1.16″	
July 15	102	40	2.23″	
July 16	96	41	0.98″	
July 17	97	41	1.96″	
July 18	98	38	1.86″	

East Tawas *(continued)*

	RECORD HIGH	RECORD LOW	RECORD PRECIPITATION	RECORD SNOWFALL
July 19	97	42	0.93″	
July 20	97	33	0.99″	
July 21	101	42	0.80″	
July 22	98	43	2.05″	
July 23	94	41	3.33″	
July 24	103	43	1.75″	
July 25	98	42	0.70″	
July 26	98	43	1.69″	
July 27	94	39	1.65″	
July 28	95	44	2.40″	
July 29	97	38	1.01″	
July 30	96	39	3.07″	
July 31	95	38	1.75″	
August 1	98	36	0.95″	
August 2	98	39	1.04″	
August 3	95	37	3.00″	
August 4	98	43	2.08″	
August 5	98	40	2.00″	
August 6	96	36	1.45″	
August 7	99	37	2.50″	
August 8	94	40	2.12″	
August 9	100	35	1.80″	
August 10	97	39	0.90″	
August 11	95	40	1.91″	
August 12	91	40	1.16″	
August 13	98	39	1.08″	
August 14	95	42	0.86″	
August 15	93	38	2.20″	
August 16	93	36	**3.72**″	
August 17	92	40	1.80″	
August 18	90	39	2.34″	
August 19	91	37	3.19″	
August 20	95	33	1.84″	
August 21	90	N/A	2.08″	
August 22	95	38	1.70″	
August 23	92	33	1.60″	
August 24	102	36	0.85″	
August 25	103	36	2.16″	
August 26	95	35	1.43″	
August 27	94	37	1.25″	
August 28	95	34	1.15″	
August 29	94	31	1.57″	
August 30	95	29	2.53″	
August 31	95	34	0.82″	
September 1	100	33	1.30″	
September 2	95	35	2.07″	
September 3	93	31	1.31″	
September 4	90	35	1.00″	
September 5	89	31	2.31″	
September 6	92	33	2.42″	

	RECORD HIGH	RECORD LOW	RECORD PRECIPITATION	RECORD SNOWFALL
September 7	95	34	1.16″	
September 8	95	34	1.66″	
September 9	95	34	0.92″	
September 10	89	34	1.42″	
September 11	101	31	1.82″	
September 12	94	30	2.00″	
September 13	86	29	1.10″	0.2″
September 14	89	30	2.20″	
September 15	93	32	1.32″	
September 16	92	29	1.64″	
September 17	89	29	2.25″	
September 18	90	27	1.52″	
September 19	92	27	0.93″	
September 20	88	27	1.48″	
September 21	87	31	1.53″	
September 22	90	29	1.33″	
September 23	90	27	1.19″	
September 24	90	26	1.45″	
September 25	87	28	0.88″	
September 26	87	25	2.62″	
September 27	89	22	1.43″	
September 28	79	21	1.29″	
September 29	84	24	1.64″	
September 30	85	22	1.13″	
October 1	80	23	0.80″	
October 2	85	26	1.18″	
October 3	80	21	2.80″	
October 4	87	25	1.00″	
October 5	84	22	3.06″	
October 6	88	19	1.10″	
October 7	90	16	2.29″	
October 8	83	23	1.11″	
October 9	93	23	1.35″	
October 10	80	22	1.32″	
October 11	86	19	1.93″	
October 12	80	20	1.05″	
October 13	83	22	1.68″	
October 14	81	19	1.65″	
October 15	82	23	1.05″	0.4″
October 16	81	25	1.10″	
October 17	84	22	1.08″	2.5″
October 18	85	16	1.14″	
October 19	83	18	1.20″	
October 20	85	21	1.48″	
October 21	84	16	0.58″	
October 22	83	19	2.01″	
October 23	81	12	1.54″	
October 24	80	19	1.80″	
October 25	78	20	1.11″	
October 26	80	17	1.30″	
October 27	78	14	1.16″	2.0″

East Tawas *(continued)*

	RECORD HIGH	RECORD LOW	RECORD PRECIPITATION	RECORD SNOWFALL
October 28	79	14	0.73″	1.0″
October 29	78	14	0.98″	
October 30	79	12	1.00″	3.5″
October 31	80	15	1.12″	
November 1	78	19	1.91″	2.5″
November 2	74	15	1.48″	1.0″
November 3	73	12	1.09″	4.0″
November 4	70	13	1.10″	1.0″
November 5	70	10	1.56″	0.5″
November 6	70	9	1.41″	1.5″
November 7	64	12	1.14″	6.0″
November 8	67	11	1.22″	2.5″
November 9	67	9	0.59″	3.0″
November 10	76	8	1.54″	1.5″
November 11	67	10	1.28″	2.5″
November 12	66	10	0.90″	3.0″
November 13	66	8	1.20″	4.0″
November 14	63	9	1.28″	6.0″
November 15	65	–4	1.34″	4.0″
November 16	69	–2	1.00″	3.5″
November 17	64	6	1.40″	6.2″
November 18	65	9	0.78″	1.4″
November 19	64	11	2.00″	3.0″
November 20	62	6	1.52″	6.7″
November 21	61	9	1.90″	2.6″
November 22	57	4	1.04″	3.0″
November 23	60	4	1.49″	3.0″
November 24	65	–1	0.79″	5.5″
November 25	57	3	0.61″	5.3″
November 26	58	4	1.10″	3.3″
November 27	53	3	1.27″	4.8″
November 28	64	2	1.13″	9.5″
November 29	58	1	0.80″	7.0″
November 30	62	1	0.99″	4.0″
December 1	59	–4	0.76″	7.5″
December 2	60	–4	1.65″	9.2″
December 3	62	–10	1.77″	6.8″
December 4	65	–1	0.99″	4.0″
December 5	57	5	0.83″	4.6″
December 6	63	–2	1.11″	4.0″
December 7	60	–2	0.96″	6.0″
December 8	57	–12	1.30″	9.0″
December 9	60	–3	0.67″	8.5″
December 10	54	–3	0.75″	6.2″
December 11	57	–4	1.25″	4.0″
December 12	54	–5	0.95″	7.9″
December 13	54	–8	1.32″	8.5″
December 14	51	–9	1.11″	9.0″
December 15	60	–6	0.94″	10.5″
December 16	52	–7	1.26″	9.6″
December 17	58	–7	1.10″	3.8″
December 18	49	–16	0.45″	5.0″
December 19	52	–9	0.80″	10.0″
December 20	53	–7	0.77″	8.0″
December 21	54	–12	0.87″	5.0″
December 22	53	–7	0.87″	4.0″
December 23	52	–9	0.82″	4.5″
December 24	49	–9	0.79″	6.2″
December 25	48	–19	1.00″	6.0″
December 26	62	–10	1.09″	7.0″
December 27	50	–14	0.90″	5.0″
December 28	52	–11	1.88″	7.0″
December 29	59	–18	0.80″	8.0″
December 30	55	–13	0.60″	6.8″
December 31	62	–10	1.07″	9.0″

Flint (Climate Record Begins 1921)

	RECORD HIGH	RECORD LOW	RECORD PRECIPITATION	RECORD SNOWFALL
January 1	50	–10	1.16″	9.6″
January 2	57	–12	1.09″	7.5″
January 3	59	–7	0.73″	4.0″
January 4	61	–16	0.79″	3.5″
January 5	59	–7	0.91″	2.4″
January 6	57	–5	0.37″	4.8″
January 7	61	–10	0.47″	1.9″
January 8	60	–10	1.12″	2.7″
January 9	54	–8	0.57″	7.0″
January 10	56	–14	0.75″	4.8″
January 11	56	–19	0.58″	6.9″
January 12	58	–11	0.72″	4.4″
January 13	60	–8	0.83″	10.4″
January 14	56	–19	0.64″	9.8″
January 15	57	–18	0.86″	2.5″
January 16	54	–14	0.45″	6.0″
January 17	60	–15	0.61″	4.3″
January 18	60	**–25**	1.34″	5.0″
January 19	54	–21	1.04″	4.5″
January 20	52	–12	1.09″	6.8″
January 21	55	–16	0.60″	5.1″
January 22	55	–13	1.17″	9.0″

Flint *(continued)*

	RECORD HIGH	RECORD LOW	RECORD PRECIPITATION	RECORD SNOWFALL
January 23	56	–13	0.33″	4.6″
January 24	57	–13	0.42″	3.4″
January 25	65	–16	0.42″	4.5″
January 26	63	–16	1.16″	**14.5**″
January 27	56	–11	0.75″	8.2″
January 28	54	–12	0.82″	2.5″
January 29	52	–11	1.32″	3.1″
January 30	51	–16	0.98″	5.5″
January 31	59	–8	1.03″	8.9″
February 1	52	–18	1.31″	6.8″
February 2	53	–22	0.54″	3.4″
February 3	53	–11	0.41″	3.8″
February 4	55	–16	0.80″	5.0″
February 5	49	–18	1.00″	5.0″
February 6	56	–5	0.66″	12.4″
February 7	50	–11	0.42″	4.2″
February 8	57	–13	0.86″	5.6″
February 9	54	–15	1.38″	2.4″
February 10	60	–19	0.58″	4.0″
February 11	68	–9	0.96″	5.2″
February 12	58	–22	1.25″	3.5″
February 13	57	–7	0.99″	3.1″
February 14	52	–8	1.60″	3.7″
February 15	62	–7	0.55″	4.6″
February 16	58	–13	2.85″	4.1″
February 17	51	–19	0.75″	3.1″
February 18	56	–11	0.70″	5.0″
February 19	59	–14	0.60″	4.1″
February 20	61	–21	0.65″	6.0″
February 21	63	–9	1.36″	4.1″
February 22	65	–5	0.95″	5.3″
February 23	63	–3	1.49″	4.3″
February 24	57	–9	0.76″	4.8″
February 25	61	–15	1.08″	10.2″
February 26	63	–10	0.58″	5.4″
February 27	58	–14	0.81″	2.2″
February 28	56	–6	0.57″	3.1″
February 29	58	–3	0.10″	2.3″
March 1	60	–11	0.37″	3.0″
March 2	58	–12	0.71″	7.5″
March 3	68	–11	1.10″	4.9″
March 4	63	–2	1.46″	5.3″
March 5	66	–10	0.40″	6.1″
March 6	67	–7	0.77″	4.5″
March 7	75	–6	0.86″	3.8″
March 8	80	–10	1.26″	5.7″
March 9	67	–5	0.87″	2.6″
March 10	65	0	1.03″	6.9″
March 11	68	–7	0.64″	3.7″
March 12	75	0	0.65″	3.8″
March 13	76	–4	0.99″	2.9″
March 14	76	5	1.01″	2.8″
March 15	78	3	0.74″	2.0″
March 16	71	6	1.92″	4.0″
March 17	73	–5	0.66″	12.6″
March 18	67	3	0.38″	3.7″
March 19	76	–1	2.33″	4.5″
March 20	78	2	0.95″	8.1″
March 21	72	7	0.92″	7.2″
March 22	82	7	1.05″	7.0″
March 23	71	2	0.69″	6.0″
March 24	73	–7	0.67″	3.6″
March 25	79	–1	1.06″	7.2″
March 26	76	4	0.98″	4.8″
March 27	78	–2	1.36″	5.0″
March 28	79	3	1.37″	2.0″
March 29	77	12	0.94″	6.0″
March 30	79	4	0.83″	4.2″
March 31	77	0	1.41″	0.7″
April 1	76	7	1.14″	6.2″
April 2	82	16	1.62″	12.2″
April 3	79	11	1.03″	8.0″
April 4	77	13	1.51″	2.6″
April 5	80	13	2.05″	5.3″
April 6	83	16	0.88″	4.0″
April 7	83	6	0.65″	1.7″
April 8	76	8	0.91″	2.3″
April 9	79	18	0.71″	2.6″
April 10	79	16	0.49″	1.5″
April 11	82	21	1.58″	1.0″
April 12	81	19	1.03″	2.1″
April 13	85	15	0.99″	1.3″
April 14	81	20	0.96″	0.5″
April 15	85	20	1.29″	0.7″
April 16	87	21	1.81″	2.5″
April 17	84	22	1.96″	6.2″
April 18	87	22	2.69″	3.0″
April 19	83	22	1.29″	1.5″
April 20	83	24	2.25″	1.0″
April 21	84	25	1.35″	2.0″
April 22	84	20	1.75″	0.6″
April 23	88	22	0.90″	3.5″
April 24	88	25	1.16″	4.3″
April 25	87	26	1.96″	1.7″
April 26	86	25	1.20″	
April 27	86	26	0.73″	3.0″
April 28	85	24	1.14″	
April 29	82	27	0.98″	
April 30	86	20	1.26″	0.8″
May 1	86	26	1.39″	0.3″

Flint *(continued)*

	RECORD HIGH	RECORD LOW	RECORD PRECIPITATION	RECORD SNOWFALL
May 2	84	27	0.75″	1.7″
May 3	88	26	0.92″	
May 4	88	23	0.65″	0.3″
May 5	90	29	0.58″	
May 6	89	27	1.57″	0.5″
May 7	88	26	1.05″	
May 8	88	28	1.63″	
May 9	89	26	2.23″	12.0″
May 10	92	22	1.50″	
May 11	90	29	1.48″	
May 12	88	29	0.78″	
May 13	89	28	0.93″	
May 14	87	32	1.53″	
May 15	90	30	1.04″	
May 16	90	30	2.16″	
May 17	89	29	1.85″	
May 18	90	29	2.23″	
May 19	89	29	1.19″	
May 20	92	32	1.69″	
May 21	93	31	1.03″	
May 22	91	32	1.12″	
May 23	88	28	1.90″	
May 24	88	32	1.18″	0.2″
May 25	88	30	2.23″	
May 26	92	32	2.00″	0.5″
May 27	90	31	0.98″	
May 28	91	31	1.46″	
May 29	92	34	0.52″	
May 30	92	33	0.92″	
May 31	93	34	0.99″	
June 1	104	33	3.48″	
June 2	102	36	1.64″	
June 3	96	35	1.67″	
June 4	98	33	1.50″	
June 5	97	36	0.52″	
June 6	100	34	1.55″	
June 7	95	40	1.20″	
June 8	98	35	1.76″	
June 9	96	36	1.61″	
June 10	95	35	1.04″	
June 11	96	35	2.24″	
June 12	93	41	0.98″	
June 13	93	42	2.52″	
June 14	97	37	1.23″	
June 15	93	41	2.20″	
June 16	96	38	1.10″	
June 17	97	41	1.35″	
June 18	97	41	1.33″	
June 19	96	44	1.70″	
June 20	97	40	0.85″	
June 21	98	40	1.80″	

	RECORD HIGH	RECORD LOW	RECORD PRECIPITATION	RECORD SNOWFALL
June 22	96	37	1.88″	
June 23	98	44	0.99″	
June 24	95	42	1.90″	
June 25	101	40	1.99″	
June 26	99	43	0.82″	
June 27	100	44	1.22″	
June 28	100	46	1.48″	
June 29	100	40	1.76″	
June 30	98	42	1.68″	
July 1	102	40	1.34″	
July 2	100	40	1.26″	
July 3	99	45	1.10″	
July 4	102	41	2.06″	
July 5	102	43	1.03″	
July 6	101	41	1.63″	
July 7	101	45	1.21″	
July 8	**108**	44	3.72″	
July 9	105	42	1.55″	
July 10	102	43	1.63″	
July 11	104	41	2.20″	
July 12	105	47	2.50″	
July 13	**108**	45	2.50″	
July 14	105	45	1.45″	
July 15	99	41	2.17″	
July 16	97	45	1.07″	
July 17	101	45	1.34″	
July 18	100	45	2.72″	
July 19	99	45	0.87″	
July 20	102	41	2.28″	
July 21	102	44	1.64″	
July 22	99	47	2.10″	
July 23	100	44	2.06″	
July 24	105	45	1.25″	
July 25	103	48	0.90″	
July 26	99	48	1.50″	
July 27	98	45	2.03″	
July 28	100	48	2.26″	
July 29	98	42	2.65″	
July 30	98	44	1.12″	
July 31	97	47	2.18″	
August 1	98	43	1.80″	
August 2	97	42	1.50″	
August 3	98	41	1.66″	
August 4	97	45	2.42″	
August 5	100	44	1.12″	
August 6	103	44	1.07″	
August 7	99	46	1.28″	
August 8	98	44	4.51″	
August 9	96	39	2.23″	
August 10	98	43	0.89″	

Flint *(continued)*

	RECORD HIGH	RECORD LOW	RECORD PRECIPITATION	RECORD SNOWFALL
August 11	96	41	1.33″	
August 12	94	39	1.22″	
August 13	96	40	0.75″	
August 14	97	41	1.52″	
August 15	95	42	0.84″	
August 16	95	43	4.45″	
August 17	96	43	3.78″	
August 18	99	42	2.23″	
August 19	96	41	2.23″	
August 20	94	41	1.34″	
August 21	97	41	4.00″	
August 22	98	40	1.22″	
August 23	98	41	2.61″	
August 24	101	40	2.64″	
August 25	95	41	1.23″	
August 26	95	43	2.90″	
August 27	98	42	1.54″	
August 28	95	39	1.55″	
August 29	96	37	1.51″	
August 30	96	38	0.85″	
August 31	97	39	1.50″	
September 1	97	40	1.13″	
September 2	97	39	1.19″	
September 3	97	35	1.66″	
September 4	91	37	3.12″	
September 5	94	38	1.01″	
September 6	93	34	2.24″	
September 7	93	38	1.10″	
September 8	97	37	2.34″	
September 9	96	38	1.20″	
September 10	93	36	**6.04″**	
September 11	95	36	1.97″	
September 12	94	34	1.04″	
September 13	93	33	2.04″	
September 14	98	33	2.63″	
September 15	100	38	0.65″	
September 16	96	33	0.81″	
September 17	94	30	1.68″	
September 18	93	34	1.72″	
September 19	90	33	1.26″	
September 20	89	33	0.79″	
September 21	90	32	1.73″	
September 22	90	30	3.62″	
September 23	89	27	2.14″	
September 24	89	31	1.10″	
September 25	86	30	2.62″	
September 26	90	30	1.79″	
September 27	86	27	1.43″	
September 28	87	26	1.04″	
September 29	84	28	1.46″	
September 30	86	30	2.24″	

	RECORD HIGH	RECORD LOW	RECORD PRECIPITATION	RECORD SNOWFALL
October 1	89	28	1.20″	
October 2	85	27	0.63″	
October 3	89	25	1.72″	
October 4	88	27	1.21″	
October 5	88	25	0.87″	
October 6	89	26	2.00″	
October 7	88	23	1.16″	
October 8	88	25	0.83″	
October 9	85	25	1.07″	
October 10	85	26	1.50″	
October 11	85	22	0.76″	
October 12	83	23	0.72″	2.3″
October 13	83	24	1.03″	
October 14	82	25	1.42″	
October 15	86	26	1.50″	
October 16	87	25	1.90″	
October 17	87	24	0.72″	
October 18	83	22	1.15″	
October 19	80	20	0.67″	3.5″
October 20	81	21	0.71″	1.8″
October 21	83	19	0.78″	0.9″
October 22	80	23	1.85″	0.5″
October 23	83	23	1.85″	0.8″
October 24	81	22	1.04″	
October 25	81	20	0.71″	
October 26	81	24	0.69″	1.3″
October 27	76	21	1.05″	2.6″
October 28	75	22	0.96″	1.0″
October 29	75	20	1.31″	
October 30	77	19	0.90″	0.5″
October 31	78	20	0.90″	0.4″
November 1	79	22	1.00″	
November 2	79	20	1.73″	4.8″
November 3	75	12	0.85″	3.8″
November 4	72	18	1.34″	1.8″
November 5	76	14	1.87″	0.3″
November 6	73	13	1.04″	10.5″
November 7	76	16	1.29″	2.9″
November 8	69	12	0.47″	2.0″
November 9	73	14	1.58″	7.0″
November 10	68	17	0.98″	0.5″
November 11	68	14	1.40″	3.0″
November 12	68	15	1.35″	1.4″
November 13	71	12	0.85″	1.2″
November 14	66	12	0.87″	3.3″
November 15	69	10	1.08″	6.0″
November 16	68	7	0.89″	4.8″
November 17	70	9	1.93″	2.9″
November 18	68	11	1.03″	1.0″
November 19	70	13	1.20″	3.0″
November 20	68	14	1.40″	4.0″

Flint *(continued)*

	RECORD HIGH	RECORD LOW	RECORD PRECIPITATION	RECORD SNOWFALL
November 21	66	7	0.65″	4.4″
November 22	65	9	1.02″	3.2″
November 23	68	10	0.61″	2.4″
November 24	64	5	1.32″	4.2″
November 25	63	9	1.71″	6.5″
November 26	63	–7	1.03″	4.0″
November 27	66	4	1.06″	4.0″
November 28	64	2	1.64″	15.0″
November 29	68	1	0.45″	3.9″
November 30	64	2	0.82″	6.3″
December 1	63	4	0.88″	7.9″
December 2	66	–5	0.61″	2.9″
December 3	67	–8	0.69″	2.2″
December 4	64	–4	0.56″	3.7″
December 5	70	6	0.85″	6.5″
December 6	66	–6	0.91″	6.9″
December 7	59	0	0.86″	7.7″
December 8	62	–4	0.80″	4.8″
December 9	59	–4	0.49″	3.5″
December 10	63	–5	0.74″	3.4″
December 11	61	–8	0.91″	10.8″
December 12	61	–8	1.54″	4.0″
December 13	55	–3	0.30″	4.7″
December 14	63	0	1.31″	7.1″
December 15	60	–3	1.74″	6.0″
December 16	62	–6	0.53″	6.6″
December 17	53	–7	0.91″	6.5″
December 18	54	–7	0.47″	4.6″
December 19	55	–11	0.69″	7.5″
December 20	59	–8	0.65″	5.6″
December 21	60	–9	1.20″	4.7″
December 22	56	–7	0.68″	5.5″
December 23	55	–12	0.80″	4.9″
December 24	56	–10	1.42″	3.4″
December 25	65	–13	0.50″	5.2″
December 26	51	–9	0.59″	3.7″
December 27	62	–9	1.38″	2.5″
December 28	63	–11	0.63″	2.4″
December 29	63	–7	1.37″	3.7″
December 30	59	–10	1.18″	4.4″
December 31	61	–11	1.04″	7.8″

Gaylord (Climate Record Begins 1893)

	RECORD HIGH	RECORD LOW	RECORD PRECIPITATION	RECORD SNOWFALL
January 1	50	–16	0.43″	8.2″
January 2	49	–16	1.20″	12.2″
January 3	51	–19	2.50″	11.0″
January 4	45	–30	0.58″	8.0″
January 5	46	–16	0.73″	5.0″
January 6	42	**–39**	0.80″	7.0″
January 7	46	–15	0.52″	11.0″
January 8	49	–21	0.39″	8.0″
January 9	47	–24	0.60″	9.5″
January 10	42	–32	0.70″	10.5″
January 11	47	–22	0.47″	8.5″
January 12	44	–29	0.52″	6.0″
January 13	46	–22	0.58″	9.0″
January 14	51	–19	1.03″	6.0″
January 15	47	–31	0.69″	9.0″
January 16	49	–21	0.40″	6.5″
January 17	48	–22	0.65″	9.0″
January 18	53	–20	0.42″	5.2″
January 19	47	–26	0.77″	5.0″
January 20	52	–15	0.82″	5.2″
January 21	50	–21	0.43″	8.0″
January 22	47	–17	0.47″	8.0″
January 23	50	–23	0.75″	6.0″
January 24	48	–22	0.43″	9.0″
January 25	47	–17	1.09″	13.0″
January 26	53	–23	1.19″	9.0″
January 27	52	–22	1.06″	8.0″
January 28	46	–21	0.60″	7.0″
January 29	45	–31	0.51″	4.5″
January 30	45	–21	0.87″	9.5″
January 31	45	–15	1.01″	10.0″
February 1	44	–15	0.70″	7.0″
February 2	42	–25	0.88″	7.5″
February 3	54	–30	1.00″	10.0″
February 4	51	–23	0.80″	8.0″
February 5	52	–16	1.00″	11.0″
February 6	45	–13	0.62″	8.0″
February 7	45	–20	0.32″	4.0″
February 8	47	–19	0.70″	8.0″
February 9	48	–27	0.60″	6.0″
February 10	49	–31	0.60″	7.0″
February 11	54	–25	0.40″	6.0″
February 12	53	–23	0.37″	5.5″
February 13	52	–23	0.67″	8.0″
February 14	46	–16	3.78″	9.0″
February 15	49	–16	0.73″	7.5″
February 16	51	–16	0.49″	6.0″

Gaylord *(continued)*

	RECORD HIGH	RECORD LOW	RECORD PRECIPITATION	RECORD SNOWFALL
February 17	55	–37	0.40″	7.0″
February 18	54	–20	0.78″	4.0″
February 19	53	–23	0.78″	6.0″
February 20	56	–23	1.23″	3.5″
February 21	52	–15	0.93″	9.0″
February 22	55	–19	0.72″	12.0″
February 23	58	–20	0.54″	8.0″
February 24	56	–21	1.21″	6.0″
February 25	54	–26	0.36″	6.5″
February 26	57	–22	1.97″	9.5″
February 27	53	–11	0.93″	11.0″
February 28	54	–12	0.81″	6.0″
February 29	55	–27	0.40″	4.0″
March 1	53	–22	0.53″	5.0″
March 2	50	–24	1.11″	7.5″
March 3	56	–21	0.58″	7.0″
March 4	58	–16	0.97″	6.5″
March 5	53	–21	1.28″	6.0″
March 6	61	–13	0.79″	12.0″
March 7	74	–21	0.70″	8.0″
March 8	76	–26	0.46″	5.0″
March 9	67	–19	1.48″	17.0″
March 10	62	–12	0.76″	8.0″
March 11	64	–18	1.50″	8.0″
March 12	65	–27	0.60″	11.0″
March 13	66	–12	1.10″	4.0″
March 14	69	–16	0.64″	7.0″
March 15	74	–10	1.41″	9.0″
March 16	69	–9	1.58″	4.4″
March 17	66	–6	1.00″	5.7″
March 18	65	–12	0.37″	5.0″
March 19	68	–7	1.00″	8.0″
March 20	63	–10	0.63″	5.5″
March 21	66	–6	0.83″	6.0″
March 22	60	–6	1.60″	13.0″
March 23	63	2	0.56″	5.5″
March 24	76	–10	1.00″	11.5″
March 25	65	–11	1.00″	4.0″
March 26	70	–19	0.58″	9.5″
March 27	70	–5	0.44″	3.2″
March 28	70	–8	0.50″	3.5″
March 29	72	–11	0.87″	4.0″
March 30	68	0	0.97″	7.5″
March 31	70	–10	2.60″	12.0″
April 1	68	–6	0.86″	5.0″
April 2	77	4	0.71″	5.0″
April 3	75	3	1.25″	7.0″
April 4	71	–1	1.48″	9.0″
April 5	76	–3	1.47″	8.0″
April 6	76	0	0.69″	5.0″
April 7	81	–4	0.94″	12.0″
April 8	74	–4	1.37″	3.5″
April 9	72	5	0.87″	3.5″
April 10	75	6	0.89″	12.0″
April 11	77	10	1.10″	6.0″
April 12	78	1	1.05″	6.0″
April 13	75	12	1.23″	3.0″
April 14	80	10	1.20″	1.0″
April 15	80	10	1.00″	8.0″
April 16	86	14	1.06″	3.0″
April 17	84	12	1.25″	4.3″
April 18	81	14	0.80″	3.0″
April 19	80	11	1.18″	4.3″
April 20	83	11	1.04″	10.0″
April 21	81	11	1.59″	5.5″
April 22	88	11	0.73″	5.5″
April 23	88	12	0.52″	5.0″
April 24	86	14	1.40″	8.0″
April 25	85	18	0.66″	5.0″
April 26	85	4	1.85″	4.5″
April 27	84	20	1.10″	3.0″
April 28	83	20	0.74″	3.0″
April 29	86	21	0.54″	5.4″
April 30	85	22	0.55″	4.5″
May 1	84	16	1.37″	4.0″
May 2	85	20	0.89″	2.5″
May 3	87	23	1.20″	3.5″
May 4	85	19	1.10″	3.0″
May 5	85	16	0.98″	4.2″
May 6	87	22	0.93″	5.6″
May 7	86	20	2.28″	0.5″
May 8	87	18	3.18″	2.0″
May 9	86	15	1.34″	1.0″
May 10	87	12	1.76″	1.0″
May 11	85	22	1.55″	3.0″
May 12	84	25	1.36″	1.0″
May 13	87	20	0.95″	1.0″
May 14	87	21	1.06″	
May 15	90	18	2.10″	
May 16	87	20	0.80″	1.0″
May 17	88	20	1.07″	1.5″
May 18	86	24	1.30″	
May 19	87	23	3.50″	
May 20	92	25	2.00″	
May 21	91	13	1.06″	
May 22	88	12	0.95″	0.9″
May 23	89	21	1.30″	0.2″
May 24	89	19	1.40″	
May 25	88	15	1.08″	
May 26	85	21	1.26″	
May 27	90	22	1.10″	6.0″

Gaylord *(continued)*

	RECORD HIGH	RECORD LOW	RECORD PRECIPITATION	RECORD SNOWFALL
May 28	91	21	2.38″	
May 29	90	26	0.84″	
May 30	93	28	1.03″	
May 31	95	22	1.45″	0.5″
June 1	96	30	1.96″	
June 2	90	30	1.28″	
June 3	88	29	1.58″	
June 4	91	26	0.63″	
June 5	90	28	1.51″	
June 6	89	22	1.05″	
June 7	92	34	0.70″	
June 8	93	26	1.25″	
June 9	89	31	1.36″	
June 10	90	29	1.02″	
June 11	91	30	1.17″	
June 12	91	25	1.12″	
June 13	95	32	1.08″	
June 14	95	30	1.58″	
June 15	92	34	1.97″	
June 16	93	33	1.06″	
June 17	92	34	0.89″	
June 18	96	33	1.35″	
June 19	93	34	1.62″	
June 20	94	31	1.25″	
June 21	94	33	1.00″	
June 22	95	31	0.99″	
June 23	95	34	1.77″	
June 24	93	32	1.16″	
June 25	94	32	1.12″	
June 26	100	36	1.95″	
June 27	100	36	1.72″	
June 28	95	34	1.32″	
June 29	94	35	3.02″	
June 30	97	32	0.94″	
July 1	**101**	36	2.49″	
July 2	100	36	0.88″	
July 3	95	38	2.00″	
July 4	95	38	2.04″	
July 5	95	39	2.15″	
July 6	99	33	1.09″	
July 7	95	39	1.61″	
July 8	94	36	1.33″	
July 9	94	39	1.00″	
July 10	92	40	1.80″	
July 11	**101**	35	1.45″	
July 12	99	40	1.17″	
July 13	94	40	1.45″	
July 14	96	40	2.67″	
July 15	95	38	1.48″	
July 16	94	39	1.78″	
July 17	92	39	1.35″	
July 18	93	39	1.32″	
July 19	96	36	2.18″	
July 20	97	38	1.98″	
July 21	95	35	2.00″	
July 22	94	37	1.89″	
July 23	92	41	1.48″	
July 24	92	39	1.96″	
July 25	91	39	1.50″	
July 26	94	36	1.29″	
July 27	94	38	1.10″	
July 28	93	30	3.04″	
July 29	98	40	0.53″	
July 30	**101**	37	1.00″	
July 31	96	37	1.15″	
August 1	96	40	1.23″	
August 2	95	32	1.70″	
August 3	94	37	1.55″	
August 4	93	30	2.42″	
August 5	94	34	1.00″	
August 6	96	36	1.25″	
August 7	94	36	1.37″	
August 8	90	39	1.35″	
August 9	96	34	2.65″	
August 10	91	39	2.00″	
August 11	90	34	1.29″	
August 12	92	33	1.66″	
August 13	92	34	1.30″	
August 14	94	36	1.23″	
August 15	92	36	1.07″	
August 16	93	29	1.18″	
August 17	97	29	**5.00″**	
August 18	94	35	1.24″	
August 19	95	36	1.87″	
August 20	97	32	2.27″	
August 21	99	37	2.75″	
August 22	95	37	2.07″	
August 23	91	34	1.78″	
August 24	91	34	2.07″	
August 25	98	32	1.40″	
August 26	91	37	1.28″	
August 27	93	36	1.43″	
August 28	94	29	1.01″	
August 29	92	26	2.50″	
August 30	92	31	2.01″	
August 31	94	34	2.77″	
September 1	96	32	2.73″	
September 2	96	31	2.44″	
September 3	94	35	1.32″	
September 4	93	29	2.21″	

Gaylord *(continued)*

	RECORD HIGH	RECORD LOW	RECORD PRECIPITATION	RECORD SNOWFALL
September 5	91	29	1.63″	
September 6	90	30	1.80″	
September 7	91	34	0.65″	
September 8	92	31	1.75″	
September 9	90	29	2.05″	
September 10	90	32	2.52″	
September 11	88	30	1.42″	
September 12	96	29	1.50″	
September 13	94	32	1.70″	
September 14	94	30	3.70″	
September 15	87	29	1.85″	
September 16	88	30	1.53″	
September 17	88	27	1.35″	
September 18	89	26	1.63″	
September 19	85	28	2.10″	
September 20	87	27	1.28″	1.2″
September 21	87	28	2.10″	
September 22	87	26	1.73″	
September 23	83	24	2.04″	
September 24	84	28	3.76″	
September 25	83	29	2.00″	0.5″
September 26	80	25	1.12″	0.5″
September 27	83	23	1.10″	
September 28	88	22	1.13″	3.5″
September 29	88	23	2.66″	
September 30	88	22	2.62″	
October 1	85	24	1.10″	2.0″
October 2	83	24	0.91″	
October 3	84	24	2.44″	0.5″
October 4	82	24	1.73″	
October 5	85	21	1.77″	
October 6	81	21	2.11″	1.5″
October 7	85	21	1.74″	3.0″
October 8	85	22	0.95″	
October 9	86	21	0.80″	2.0″
October 10	78	21	0.93″	0.5″
October 11	83	23	0.72″	2.5″
October 12	81	21	1.20″	6.0″
October 13	79	24	2.82″	3.0″
October 14	80	24	1.61″	3.0″
October 15	80	19	1.62″	4.0″
October 16	80	24	1.02″	4.2″
October 17	82	17	1.70″	5.0″
October 18	79	10	1.04″	5.5″
October 19	79	19	0.79″	2.0″
October 20	80	19	0.96″	7.0″
October 21	83	18	0.50″	4.0″
October 22	78	12	0.62″	5.0″
October 23	81	10	1.34″	3.5″
October 24	80	15	1.39″	4.0″
October 25	76	22	1.93″	3.0″
October 26	74	18	1.12″	4.0″
October 27	77	20	0.85″	3.5″
October 28	70	17	0.64″	4.0″
October 29	71	14	0.60″	6.0″
October 30	74	13	0.71″	2.0″
October 31	72	15	0.97″	3.0″
November 1	73	17	1.35″	6.0″
November 2	69	16	2.14″	5.0″
November 3	75	9	1.02″	13.0″
November 4	75	14	1.52″	7.0″
November 5	75	7	1.52″	14.0″
November 6	70	–5	1.56″	6.5″
November 7	69	4	1.95″	15.0″
November 8	63	10	1.20″	10.5″
November 9	71	12	0.97″	6.0″
November 10	68	10	1.37″	6.0″
November 11	66	4	1.00″	11.5″
November 12	63	10	2.00″	5.0″
November 13	62	3	1.24″	8.5″
November 14	61	10	1.54″	6.0″
November 15	63	–1	1.84″	7.0″
November 16	65	6	1.07″	11.0″
November 17	70	5	0.72″	6.0″
November 18	68	7	1.39″	12.0″
November 19	71	10	1.00″	10.0″
November 20	65	0	1.04″	10.0″
November 21	64	1	1.20″	9.0″
November 22	62	6	0.78″	6.0″
November 23	61	–6	2.00″	**20.0″**
November 24	57	4	0.80″	9.0″
November 25	60	5	1.05″	6.5″
November 26	61	3	1.55″	6.5″
November 27	64	–7	1.58″	7.0″
November 28	58	0	2.05″	8.5″
November 29	59	–7	1.12″	14.2″
November 30	61	5	0.83″	9.5″
December 1	58	–6	1.21″	8.5″
December 2	62	–14	0.70″	10.5″
December 3	65	–8	0.76″	7.5″
December 4	64	–5	0.86″	9.0″
December 5	62	–2	1.54″	11.0″
December 6	61	–13	0.44″	9.3″
December 7	53	–12	1.10″	7.0″
December 8	52	–6	0.66″	10.0″
December 9	52	–3	0.80″	8.0″
December 10	46	–10	1.15″	9.0″
December 11	50	–13	0.53″	10.8″
December 12	49	–13	0.74″	10.0″
December 13	48	–15	1.50″	6.5″
December 14	57	–8	0.63″	8.0″

Gaylord *(continued)*

	RECORD HIGH	RECORD LOW	RECORD PRECIPITATION	RECORD SNOWFALL
December 15	51	–6	1.48″	15.0″
December 16	47	–12	0.46″	9.0″
December 17	42	–10	0.90″	7.5″
December 18	39	–22	0.41″	6.0″
December 19	50	–20	0.45″	6.2″
December 20	45	–27	0.80″	7.0″
December 21	52	–12	0.48″	9.5″
December 22	50	–12	0.86″	10.5″
December 23	50	–8	0.43″	7.0″
December 24	44	–11	0.48″	6.0″
December 25	59	–20	0.51″	7.5″
December 26	55	–19	0.80″	4.5″
December 27	46	–19	0.70″	11.7″
December 28	52	–10	1.20″	11.0″
December 29	44	–22	0.50″	8.0″
December 30	41	–16	0.78″	10.5″
December 31	57	–20	0.61″	6.2″

Gladwin (Climate Record Begins 1892)

	RECORD HIGH	RECORD LOW	RECORD PRECIPITATION	RECORD SNOWFALL
January 1	57	–15	0.95″	10.0″
January 2	54	–13	3.20″	4.5″
January 3	55	–12	1.10″	9.0″
January 4	55	–23	0.91″	3.0″
January 5	52	–16	1.76″	6.0″
January 6	52	–15	0.90″	6.0″
January 7	47	–13	0.30″	3.6″
January 8	55	–16	0.95″	8.0″
January 9	50	–21	0.69″	4.0″
January 10	51	–19	1.10″	7.0″
January 11	51	–26	0.51″	7.0″
January 12	46	–24	0.96″	11.0″
January 13	51	–28	0.95″	5.1″
January 14	48	–14	1.35″	6.0″
January 15	51	–16	0.43″	7.0″
January 16	51	–21	0.44″	4.3″
January 17	49	–21	1.10″	5.0″
January 18	48	–21	1.01″	7.0″
January 19	57	–21	0.80″	4.5″
January 20	45	–23	0.72″	7.0″
January 21	51	–23	1.30″	6.0″
January 22	52	–16	0.85″	7.9″
January 23	51	–17	0.75″	5.0″
January 24	50	–14	1.50″	4.0″
January 25	57	–20	1.37″	8.8″
January 26	62	–22	1.02″	6.0″
January 27	50	–25	1.60″	13.0″
January 28	51	–19	1.30″	7.0″
January 29	48	–16	0.93″	5.5″
January 30	49	–22	0.91″	8.0″
January 31	49	–26	0.37″	6.0″
February 1	52	–20	0.95″	4.0″
February 2	56	–24	0.32″	5.0″
February 3	46	–27	0.35″	3.0″
February 4	51	–25	0.32″	4.8″
February 5	49	–28	0.80″	4.0″
February 6	52	–21	0.65″	4.5″
February 7	45	–24	0.79″	8.0″
February 8	59	–19	1.70'	6.8″
February 9	53	–30	0.87″	5.0″
February 10	48	–30	1.33″	7.7″
February 11	58	–31	1.00″	5.0″
February 12	63	–30	0.71″	7.0″
February 13	46	–23	1.40″	3.0″
February 14	49	–20	0.83″	5.0″
February 15	50	–22	0.44″	4.0″
February 16	49	–25	0.44″	3.8″
February 17	50	–27	1.25″	3.0″
February 18	48	–25	0.55″	2.0″
February 19	54	–20	1.28″	7.0″
February 20	54	**–39**	0.73″	4.0″
February 21	58	–25	0.75″	7.0″
February 22	59	–28	1.50″	12.0″
February 23	58	–10	0.82″	10.0″
February 24	63	–22	0.68″	6.5″
February 25	52	–21	1.55″	5.5″
February 26	55	–16	0.90″	8.5″
February 27	62	–19	1.18″	3.3″
February 28	52	–10	1.08″	6.0″
February 29	51	–16	0.41″	4.0″
March 1	60	–14	0.55″	6.0″
March 2	59	–14	1.05″	8.0″
March 3	57	–15	1.08″	12.0″
March 4	65	–14	1.00″	8.0″
March 5	57	–10	2.30″	5.0″
March 6	58	–13	1.07″	6.0″
March 7	62	–12	1.20″	4.0″
March 8	74	–14	0.67″	6.2″
March 9	78	–9	1.00″	4.0″
March 10	60	–15	1.35″	12.0″

Gladwin *(continued)*

	RECORD HIGH	RECORD LOW	RECORD PRECIPITATION	RECORD SNOWFALL
March 11	64	–5	1.00″	3.5″
March 12	68	–12	1.20″	3.0″
March 13	70	–7	1.50″	6.0″
March 14	70	–6	0.80″	8.0″
March 15	75	–6	0.99″	3.5″
March 16	79	–14	1.30″	2.0″
March 17	72	–6	1.28″	7.5″
March 18	71	–5	2.25″	6.0″
March 19	74	–8	1.10″	12.0″
March 20	66	–8	0.72″	4.0″
March 21	71	–8	1.07″	4.0″
March 22	76	–7	1.14″	7.5″
March 23	74	0	1.50″	5.0″
March 24	78	–3	2.10″	5.0″
March 25	76	–10	0.84″	6.0″
March 26	74	4	0.99″	8.0″
March 27	79	2	1.00″	6.0″
March 28	79	6	1.73″	2.0″
March 29	82	9	0.87″	3.0″
March 30	75	7	1.24″	4.0″
March 31	75	2	1.04″	6.0″
April 1	73	10	1.45″	6.0″
April 2	81	5	1.00″	4.5″
April 3	76	7	1.20″	6.0″
April 4	82	5	0.98″	3.0″
April 5	77	8	2.70″	6.7″
April 6	78	10	1.37″	2.0″
April 7	81	10	1.76″	3.0″
April 8	82	12	0.82″	5.0″
April 9	78	1	1.75″	2.0″
April 10	82	15	2.50″	5.0″
April 11	81	12	1.25″	0.3″
April 12	84	11	0.88″	4.0″
April 13	77	10	1.00″	3.5″
April 14	76	16	1.50″	1.0″
April 15	81	15	1.80″	12.0″
April 16	84	10	1.58″	2.3″
April 17	88	18	1.38″	4.0″
April 18	85	16	1.96″	2.0″
April 19	88	17	1.00″	1.2″
April 20	83	12	1.16″	2.0″
April 21	84	21	1.25″	1.2″
April 22	85	19	0.70″	2.0″
April 23	86	18	1.28″	3.0″
April 24	85	24	0.93″	1.5″
April 25	84	20	1.48″	2.2″
April 26	90	22	1.15″	3.0″
April 27	87	22	1.26″	
April 28	88	20	1.52″	0.4″
April 29	87	24	2.10″	
April 30	84	22	1.50″	

	RECORD HIGH	RECORD LOW	RECORD PRECIPITATION	RECORD SNOWFALL
May 1	85	21	1.31″	0.8″
May 2	89	22	1.23″	0.3″
May 3	88	21	1.60″	2.0″
May 4	88	23	1.14″	0.3″
May 5	91	22	0.92″	
May 6	88	25	1.33″	
May 7	86	25	1.28″	
May 8	88	21	1.40″	
May 9	89	24	2.22″	
May 10	88	21	1.80″	
May 11	91	23	1.62″	
May 12	87	24	2.15″	
May 13	91	23	0.97″	
May 14	91	26	1.10″	6.0″
May 15	90	26	1.40″	
May 16	91	26	2.20″	
May 17	89	27	1.10″	
May 18	90	24	1.37″	
May 19	89	27	1.70″	
May 20	91	27	**5.00**″	
May 21	91	30	2.20″	
May 22	91	26	1.31″	
May 23	89	25	1.52″	
May 24	90	28	1.91″	
May 25	89	29	1.86″	
May 26	89	28	1.80″	
May 27	88	26	1.36″	
May 28	92	30	1.77″	
May 29	91	27	1.10″	
May 30	95	31	1.04″	
May 31	90	34	1.96″	
June 1	99	32	3.05″	
June 2	98	33	1.25″	
June 3	93	33	0.92″	
June 4	95	32	1.00″	
June 5	92	33	1.50″	
June 6	94	34	1.00″	
June 7	96	32	1.75″	
June 8	94	30	1.55″	
June 9	95	34	1.75″	
June 10	97	32	0.76″	
June 11	96	30	1.07″	
June 12	95	39	0.70″	
June 13	92	35	1.03″	
June 14	93	34	2.40″	
June 15	96	39	1.34″	
June 16	97	36	2.02″	
June 17	98	36	1.49″	
June 18	97	40	2.65″	
June 19	98	37	3.32″	
June 20	101	36	1.48″	

Gladwin *(continued)*

	RECORD HIGH	RECORD LOW	RECORD PRECIPITATION	RECORD SNOWFALL
June 21	98	30	2.15″	
June 22	96	35	1.85″	
June 23	97	40	2.83″	
June 24	95	40	1.30″	
June 25	97	37	2.12″	
June 26	98	40	1.17″	
June 27	99	34	1.73″	
June 28	102	40	1.51″	
June 29	98	39	1.20″	
June 30	101	37	1.75″	
July 1	101	37	0.97″	
July 2	102	38	0.52″	
July 3	97	34	0.80″	
July 4	98	41	2.40″	
July 5	98	37	2.00″	
July 6	100	38	1.15″	
July 7	101	40	1.75″	
July 8	101	43	2.22″	
July 9	101	41	1.69″	
July 10	99	38	1.45″	
July 11	102	34	2.38″	
July 12	103	38	1.01″	
July 13	**105**	36	1.69″	
July 14	97	40	1.19″	
July 15	101	42	3.97″	
July 16	96	41	2.00″	
July 17	98	38	1.22″	
July 18	99	41	2.10″	
July 19	96	43	3.71″	
July 20	97	40	3.56″	
July 21	98	38	2.20″	
July 22	95	45	0.85″	
July 23	95	41	1.85″	
July 24	99	43	1.26″	
July 25	99	40	0.74″	
July 26	100	41	1.92″	
July 27	100	41	1.83″	
July 28	96	41	1.60″	
July 29	99	42	1.60″	
July 30	94	41	1.40″	
July 31	94	36	1.67″	
August 1	99	40	3.33″	
August 2	96	37	1.33″	
August 3	97	39	1.30″	
August 4	96	40	1.85″	
August 5	98	39	1.54″	
August 6	99	43	1.02″	
August 7	96	41	1.40″	
August 8	98	39	4.25″	
August 9	100	39	1.62″	

	RECORD HIGH	RECORD LOW	RECORD PRECIPITATION	RECORD SNOWFALL
August 10	97	40	1.20″	
August 11	94	43	1.33″	
August 12	91	39	1.40″	
August 13	95	38	1.53″	
August 14	97	39	1.54″	
August 15	95	39	2.14″	
August 16	90	35	2.02″	
August 17	94	41	1.42″	
August 18	97	39	1.40″	
August 19	98	35	3.10″	
August 20	97	36	1.55″	
August 21	100	31	2.38″	
August 22	95	37	1.20″	
August 23	97	37	1.65″	
August 24	98	35	1.14″	
August 25	96	38	1.85″	
August 26	95	37	2.00″	
August 27	95	40	1.25″	
August 28	97	31	1.46″	
August 29	97	36	1.75″	
August 30	96	34	1.36″	
August 31	94	36	2.39″	
September 1	97	30	1.21″	
September 2	98	28	2.10″	
September 3	94	26	0.75″	
September 4	94	30	2.09″	
September 5	90	33	2.35″	
September 6	90	31	1.01″	
September 7	93	32	1.06″	
September 8	93	33	2.32″	
September 9	94	34	1.99″	
September 10	92	34	1.45″	
September 11	95	31	4.30″	
September 12	92	29	1.93″	
September 13	92	30	1.31″	
September 14	100	29	3.89″	
September 15	90	30	1.42″	
September 16	88	29	1.05″	
September 17	89	32	2.00″	
September 18	89	31	1.63″	
September 19	88	27	1.20″	
September 20	90	29	1.85″	
September 21	87	25	2.40″	
September 22	87	22	1.98″	
September 23	87	24	1.59″	
September 24	85	28	1.60″	
September 25	89	25	1.30″	
September 26	82	24	3.30″	
September 27	86	25	1.35″	
September 28	93	25	2.13″	
September 29	87	22	1.87″	

Gladwin *(continued)*

	RECORD HIGH	RECORD LOW	RECORD PRECIPITATION	RECORD SNOWFALL
September 30	85	26	2.00″	
October 1	82	19	1.06″	
October 2	86	25	1.07″	
October 3	85	24	2.98″	
October 4	85	26	1.90″	
October 5	85	23	3.03″	
October 6	87	20	2.40″	
October 7	87	17	1.83″	
October 8	89	23	1.10″	
October 9	90	23	1.96″	
October 10	85	20	1.85″	
October 11	82	18	1.10″	
October 12	83	23	0.63″	0.1″
October 13	83	21	1.56″	2.5″
October 14	85	23	1.72″	
October 15	87	15	0.76″	
October 16	84	23	1.69″	0.9″
October 17	83	11	0.80″	0.8″
October 18	83	15	1.00″	0.6″
October 19	82	19	1.15″	1.5″
October 20	80	21	1.05″	
October 21	82	18	1.47″	0.7″
October 22	82	23	1.24″	
October 23	84	16	2.56″	
October 24	82	18	1.27″	4.0″
October 25	76	18	1.25″	
October 26	75	20	1.25″	3.0″
October 27	76	13	1.21″	6.0″
October 28	73	17	1.23″	0.4″
October 29	75	15	1.50″	
October 30	86	21	1.10″	
October 31	78	17	1.00″	
November 1	77	15	1.09″	1.9″
November 2	74	15	1.44″	1.1″
November 3	73	10	1.16″	4.0″
November 4	74	10	1.34″	1.7″
November 5	72	4	1.30″	3.0″
November 6	74	6	1.53″	2.3″
November 7	72	10	1.17″	3.1″
November 8	67	7	1.00″	0.6″
November 9	72	12	2.17″	5.0″
November 10	74	14	1.40″	6.0″
November 11	70	10	1.20″	2.0″
November 12	65	12	1.05″	2.5″
November 13	65	9	1.37″	6.0″
November 14	64	8	1.65″	2.5″
November 15	68	3	1.58″	1.5″
November 16	68	3	1.35″	5.5″
November 17	69	4	1.08″	4.0″
November 18	67	4	0.79″	2.0″
November 19	70	8	1.03″	8.0″
November 20	68	6	1.03″	7.0″
November 21	65	5	0.66″	2.0″
November 22	61	5	0.92″	3.0″
November 23	69	2	1.14″	3.0″
November 24	66	–8	2.40″	3.7″
November 25	60	–2	1.40″	6.0″
November 26	62	–3	1.80″	4.9″
November 27	60	1	1.30″	3.0″
November 28	65	–4	1.50″	8.0″
November 29	60	–1	0.57″	3.7″
November 30	68	1	1.50″	3.5″
December 1	63	–3	0.83″	6.0″
December 2	58	–3	1.15″	6.0″
December 3	63	–10	1.61″	8.0″
December 4	63	–4	0.52″	6.0″
December 5	57	–1	1.10″	3.8″
December 6	69	–14	0.95″	7.0″
December 7	59	–7	1.40″	10.0″
December 8	57	–4	1.30″	4.0″
December 9	60	–10	0.70″	8.0″
December 10	58	–14	2.17″	6.0″
December 11	63	–19	1.20″	**15.0″**
December 12	56	–7	1.52″	9.0″
December 13	55	–8	1.60″	3.0″
December 14	59	–12	0.70″	4.5″
December 15	58	–14	1.74″	7.0″
December 16	54	–14	1.04″	6.0″
December 17	57	–12	3.92″	10.0″
December 18	50	–10	0.58″	3.4″
December 19	50	–19	0.50″	4.0″
December 20	55	–8	4.00″	8.6″
December 21	58	–8	0.80″	6.0″
December 22	54	–10	0.65″	6.0″
December 23	51	–9	0.68″	3.0″
December 24	50	–16	0.91″	5.0″
December 25	55	–14	0.61″	4.0″
December 26	63	–11	0.63″	4.0″
December 27	49	–14	1.26″	11.0″
December 28	55	–8	1.58″	7.4″
December 29	59	–20	0.67″	5.0″
December 30	54	–20	1.42″	12.0″
December 31	59	–24	0.46″	4.9″

Grand Rapids (Climate Record Begins 1892)

	RECORD HIGH	RECORD LOW	RECORD PRECIPITATION	RECORD SNOWFALL
January 1	66	–15	0.65″	5.7″
January 2	59	–2	1.35″	6.4″
January 3	59	–11	1.50″	9.1″
January 4	60	–20	2.15″	6.3″
January 5	55	–9	0.78″	6.4″
January 6	57	–12	0.96″	11.0″
January 7	63	–15	0.59″	5.4″
January 8	59	–13	0.74″	5.0″
January 9	57	–14	0.89″	8.0″
January 10	57	–19	1.69″	5.2″
January 11	55	–21	0.51″	6.9″
January 12	55	–14	1.85″	10.0″
January 13	58	–2	1.17″	10.5″
January 14	53	–6	0.89″	7.2″
January 15	57	–15	0.54″	5.8″
January 16	58	–16	0.60″	5.5″
January 17	56	–10	0.97″	4.7″
January 18	60	–15	1.20″	8.0″
January 19	56	–22	1.84″	6.4″
January 20	64	–15	1.36″	6.8″
January 21	61	–15	0.99″	6.1″
January 22	68	–6	1.70″	12.3″
January 23	56	–19	0.80″	8.0″
January 24	62	–19	0.88″	6.5″
January 25	66	–14	0.72″	8.4″
January 26	62	–9	1.09″	**16.1**″
January 27	62	–12	0.76″	12.2″
January 28	53	–8	0.79″	6.0″
January 29	59	–16	0.94″	7.5″
January 30	52	–22	1.03″	5.5″
January 31	57	–20	1.39″	14.0″
February 1	51	–20	1.32″	5.7″
February 2	48	–11	0.49″	5.2″
February 3	54	–17	0.61″	11.3″
February 4	53	–17	1.12″	10.0″
February 5	52	–16	1.21″	4.4″
February 6	54	–9	0.71″	8.1″
February 7	52	–15	0.46″	8.1″
February 8	60	–12	0.75″	5.4″
February 9	55	–16	1.02″	7.0″
February 10	61	–21	0.99″	6.7″
February 11	69	–21	0.79″	8.8″
February 12	55	–23	1.70″	4.2″
February 13	58	**–24**	0.68″	5.0″
February 14	53	**–24**	0.90″	7.8″
February 15	61	–7	0.43″	4.8″
February 16	59	–12	1.84″	2.8″
February 17	55	–19	0.86″	4.7″
February 18	55	–10	0.80″	7.0″
February 19	59	–15	0.85″	5.0″

	RECORD HIGH	RECORD LOW	RECORD PRECIPITATION	RECORD SNOWFALL
February 20	61	–13	1.40″	5.0″
February 21	66	–8	2.96″	6.0″
February 22	67	–7	1.06″	4.8″
February 23	63	–4	1.48″	5.9″
February 24	57	–9	1.09″	6.8″
February 25	61	–14	1.43″	8.1″
February 26	64	–9	1.37″	4.5″
February 27	67	–12	0.81″	2.5″
February 28	54	–2	1.30″	9.0″
February 29	58	–2	0.27″	4.1″
March 1	58	–5	0.97″	2.3″
March 2	60	–8	1.29″	13.6″
March 3	72	–9	1.32″	5.0″
March 4	66	–3	1.36″	8.0″
March 5	68	–8	0.82″	7.4″
March 6	68	0	0.68″	5.6″
March 7	72	–3	1.37″	5.2″
March 8	78	–13	1.02″	4.5″
March 9	68	–7	1.35″	8.2″
March 10	69	3	0.68″	6.2″
March 11	71	1	0.75″	6.7″
March 12	76	2	2.35″	3.2″
March 13	75	5	1.06″	3.6″
March 14	75	0	0.79″	10.5″
March 15	75	5	1.31″	5.2″
March 16	71	3	1.50″	10.0″
March 17	73	–4	1.47″	7.5″
March 18	75	5	1.06″	5.7″
March 19	75	0	2.12″	3.3″
March 20	74	3	0.63″	8.4″
March 21	76	9	0.79″	6.6″
March 22	82	6	0.79″	6.3″
March 23	74	4	1.38″	8.4″
March 24	78	–3	1.07″	5.3″
March 25	77	–1	2.82″	7.7″
March 26	77	6	1.20″	10.2″
March 27	81	0	1.03″	6.9″
March 28	78	8	1.02″	3.3″
March 29	82	6	0.83″	7.5″
March 30	76	1	1.21″	3.0″
March 31	78	6	1.17″	3.7″
April 1	77	13	1.34″	4.8″
April 2	79	11	1.16″	7.8″
April 3	80	10	1.16″	6.0″
April 4	77	13	1.27″	2.5″
April 5	81	12	1.87″	8.6″
April 6	81	18	1.02″	3.5″
April 7	81	3	1.06″	5.0″
April 8	74	7	1.47″	4.3″

Grand Rapids *(continued)*

	RECORD HIGH	RECORD LOW	RECORD PRECIPITATION	RECORD SNOWFALL
April 9	79	18	1.38″	4.4″
April 10	84	15	1.61″	2.0″
April 11	82	14	1.21″	4.6″
April 12	82	19	1.10″	1.2″
April 13	80	16	1.09″	2.8″
April 14	80	21	1.07″	1.4″
April 15	84	18	1.40″	4.0″
April 16	86	21	1.97″	11.8″
April 17	84	24	2.94″	1.3″
April 18	86	22	1.44″	3.0″
April 19	82	19	1.89″	1.0″
April 20	83	13	2.45″	0.4″
April 21	84	22	1.71″	2.0″
April 22	86	24	1.72″	0.1″
April 23	84	20	1.44″	0.7″
April 24	86	26	1.46″	2.4″
April 25	86	22	1.40″	3.0″
April 26	88	24	1.48″	
April 27	85	27	1.90″	0.2″
April 28	83	24	1.18″	0.9″
April 29	90	28	2.53″	2.2″
April 30	87	26	1.85″	3.6″
May 1	88	21	2.02″	0.3″
May 2	86	22	0.96″	3.0″
May 3	87	26	2.03″	1.3″
May 4	88	26	1.06″	0.4″
May 5	89	28	1.73″	0.3″
May 6	90	25	1.22″	
May 7	89	25	1.70″	
May 8	89	27	1.50″	
May 9	88	23	4.10″	5.5″
May 10	89	22	3.53″	5.5″
May 11	89	28	2.98″	
May 12	86	27	1.43″	
May 13	86	27	1.92″	
May 14	89	31	1.26″	
May 15	91	34	4.15″	
May 16	89	29	1.97″	
May 17	91	30	1.16″	
May 18	90	28	3.31″	0.2″
May 19	91	30	2.88″	
May 20	92	30	3.39″	
May 21	91	31	2.46″	
May 22	91	34	1.10″	
May 23	90	30	2.26″	
May 24	89	33	1.25″	
May 25	91	31	1.27″	
May 26	89	34	1.19″	
May 27	92	30	1.12″	
May 28	92	33	1.88″	
May 29	92	35	0.80″	

	RECORD HIGH	RECORD LOW	RECORD PRECIPITATION	RECORD SNOWFALL
May 30	92	34	1.91″	
May 31	95	36	1.37″	
June 1	102	36	2.65″	
June 2	95	35	1.42″	
June 3	96	38	1.76″	
June 4	93	32	1.09″	
June 5	95	36	**4.22**″	
June 6	95	37	1.82″	
June 7	91	38	1.87″	
June 8	97	38	2.32″	
June 9	93	37	1.77″	
June 10	93	36	1.37″	
June 11	93	33	2.64″	
June 12	96	38	1.89″	
June 13	95	39	2.23″	
June 14	96	38	2.89″	
June 15	97	43	2.41″	
June 16	96	39	2.23″	
June 17	95	40	2.01″	
June 18	96	43	1.52″	
June 19	98	45	3.15″	
June 20	102	41	1.84″	
June 21	98	40	3.36″	
June 22	97	39	1.92″	
June 23	98	38	3.17″	
June 24	96	41	2.88″	
June 25	97	41	1.81″	
June 26	94	43	2.78″	
June 27	97	45	1.30″	
June 28	97	43	2.33″	
June 29	97	43	1.47″	
June 30	99	40	1.69″	
July 1	101	44	1.43″	
July 2	98	43	3.18″	
July 3	97	43	2.85″	
July 4	100	45	1.74″	
July 5	100	44	3.56″	
July 6	100	41	2.02″	
July 7	98	45	1.66″	
July 8	101	46	3.38″	
July 9	101	48	1.34″	
July 10	102	45	1.48″	
July 11	99	43	2.72″	
July 12	106	42	1.24″	
July 13	**108**	46	1.43″	
July 14	102	45	1.22″	
July 15	95	47	2.17″	
July 16	97	42	1.74″	
July 17	97	48	1.86″	
July 18	101	47	2.15″	

Grand Rapids *(continued)*

	RECORD HIGH	RECORD LOW	RECORD PRECIPITATION	RECORD SNOWFALL
July 19	96	47	1.80″	
July 20	99	44	2.78″	
July 21	104	43	2.40″	
July 22	97	43	2.36″	
July 23	101	45	1.02″	
July 24	103	47	0.92″	
July 25	100	48	1.02″	
July 26	100	46	1.64″	
July 27	101	47	2.96″	
July 28	100	46	1.12″	
July 29	103	47	1.74″	
July 30	101	50	1.68″	
July 31	98	46	1.59″	
August 1	98	48	1.96″	
August 2	98	46	1.86″	
August 3	100	46	1.70″	
August 4	96	42	2.12″	
August 5	100	44	0.94″	
August 6	102	46	1.40″	
August 7	97	46	0.97″	
August 8	96	43	1.57″	
August 9	97	43	1.25″	
August 10	98	45	1.42″	
August 11	96	43	1.31″	
August 12	95	45	1.27″	
August 13	97	44	1.80″	
August 14	96	41	3.61″	
August 15	97	42	1.35″	
August 16	99	40	1.71″	
August 17	98	45	1.94″	
August 18	97	41	1.73″	
August 19	98	43	**4.22**″	
August 20	97	43	2.34″	
August 21	97	44	2.05″	
August 22	98	44	1.25″	
August 23	96	43	1.91″	
August 24	98	40	1.56″	
August 25	94	41	2.02″	
August 26	95	42	2.07″	
August 27	95	42	2.08″	
August 28	94	41	1.88″	
August 29	95	41	1.60″	
August 30	95	39	2.00″	
August 31	97	43	1.02″	
September 1	97	40	1.06″	
September 2	98	37	2.17″	
September 3	95	32	1.42″	
September 4	94	38	2.82″	
September 5	92	39	2.37″	
September 6	97	36	1.67″	
September 7	96	38	1.64″	
September 8	95	36	2.27″	
September 9	95	37	2.25″	
September 10	94	36	2.76″	
September 11	94	32	3.21″	
September 12	95	37	1.54″	
September 13	95	35	2.46″	
September 14	95	31	2.15″	
September 15	97	36	1.32″	
September 16	98	32	2.26″	
September 17	92	37	2.29″	
September 18	94	35	0.86″	
September 19	91	36	1.73″	
September 20	90	33	1.75″	
September 21	92	32	3.16″	
September 22	90	33	3.14″	
September 23	90	28	2.13″	
September 24	91	29	1.39″	
September 25	90	29	3.52″	
September 26	89	30	2.27″	
September 27	86	29	1.80″	
September 28	86	27	1.96″	
September 29	93	29	2.22″	
September 30	85	30	3.10″	
October 1	89	27	2.10″	
October 2	87	27	1.75″	
October 3	85	23	3.59″	
October 4	87	30	1.28″	
October 5	85	29	1.46″	
October 6	87	28	1.29″	
October 7	87	25	1.27″	0.1″
October 8	88	28	0.96″	
October 9	84	23	2.54″	
October 10	85	26	1.44″	
October 11	85	25	1.49″	
October 12	84	27	1.52″	2.0″
October 13	87	24	1.35″	0.1″
October 14	83	29	1.70″	
October 15	85	23	0.90″	
October 16	83	25	1.78″	
October 17	84	22	1.91″	
October 18	82	19	1.32″	0.9″
October 19	80	19	0.77″	4.5″
October 20	83	22	1.11″	2.2″
October 21	85	18	2.13″	1.1″
October 22	82	23	1.40″	0.9″
October 23	83	19	1.92″	1.0″
October 24	80	24	2.24″	0.1″
October 25	76	20	0.73″	2.4″
October 26	79	24	0.80″	2.0″
October 27	78	19	1.32″	8.2″

Grand Rapids *(continued)*

	RECORD HIGH	RECORD LOW	RECORD PRECIPITATION	RECORD SNOWFALL
October 28	74	23	0.84″	3.5″
October 29	77	20	1.02″	1.6″
October 30	79	18	2.83″	1.5″
October 31	79	20	1.27″	1.5″
November 1	81	21	1.24″	4.0″
November 2	77	18	2.69″	1.5″
November 3	76	17	2.17″	10.4″
November 4	73	15	1.50″	5.8″
November 5	75	6	2.66″	2.6″
November 6	77	9	1.29″	4.0″
November 7	69	14	1.01″	6.0″
November 8	72	16	0.76″	7.5″
November 9	73	16	2.53″	4.0″
November 10	67	16	1.37″	3.5″
November 11	73	18	1.19″	5.8″
November 12	68	15	0.89″	3.6″
November 13	70	10	1.79″	3.2″
November 14	68	13	1.41″	4.5″
November 15	68	10	1.84″	4.1″
November 16	68	11	1.34″	5.0″
November 17	70	9	1.59″	7.1″
November 18	70	11	1.68″	2.9″
November 19	74	9	1.33″	4.3″
November 20	71	13	1.27″	11.5″
November 21	70	14	1.31″	4.0″
November 22	66	13	0.70″	7.0″
November 23	70	2	0.88″	3.4″
November 24	66	–9	1.09″	9.7″
November 25	65	–10	1.39″	3.5″
November 26	65	3	1.16″	5.7″
November 27	65	5	2.94″	3.8″
November 28	67	5	1.58″	7.9″
November 29	65	6	1.13″	4.5″
November 30	64	6	0.65″	6.8″

	RECORD HIGH	RECORD LOW	RECORD PRECIPITATION	RECORD SNOWFALL
December 1	65	6	0.86″	4.1″
December 2	67	–11	1.48″	4.8″
December 3	64	–6	1.37″	4.0″
December 4	63	6	1.27″	5.9″
December 5	69	4	1.23″	7.7″
December 6	63	0	0.93″	7.0″
December 7	62	–4	1.12″	7.7″
December 8	64	2	0.78″	4.5″
December 9	59	–1	0.81″	4.6″
December 10	62	5	2.13″	7.2″
December 11	63	–3	1.12″	14.2″
December 12	60	–5	1.36″	6.0″
December 13	57	–6	0.81″	6.8″
December 14	59	–1	1.32″	5.1″
December 15	60	–7	2.25″	6.7″
December 16	61	–2	0.97″	5.7″
December 17	54	–5	1.37″	5.4″
December 18	55	–5	1.03″	4.0″
December 19	55	–18	0.99″	10.2″
December 20	58	–3	3.82″	7.2″
December 21	60	–8	1.26″	7.9″
December 22	57	–11	1.07″	8.5″
December 23	57	–11	1.02″	5.9″
December 24	58	–9	1.79″	7.5″
December 25	65	–3	1.01″	7.0″
December 26	50	–6	0.66″	11.9″
December 27	60	–5	1.51″	9.6″
December 28	65	–9	0.45″	5.4″
December 29	65	–4	1.38″	5.3″
December 30	60	–11	1.75″	7.0″
December 31	60	–14	0.75″	9.7″

Hancock (Climate Record Begins 1887)

	RECORD HIGH	RECORD LOW	RECORD PRECIPITATION	RECORD SNOWFALL
January 1	46	–21	0.70″	9.0″
January 2	44	–21	0.72″	9.0″
January 3	43	–9	1.20″	23.0″
January 4	41	–12	0.61″	10.5″
January 5	37	–15	0.72″	8.5″
January 6	41	–14	0.76″	13.1″
January 7	42	–19	0.88″	12.0″
January 8	42	–19	0.71″	14.0″
January 9	40	–20	0.85″	11.0″
January 10	42	–13	0.68″	9.7″

	RECORD HIGH	RECORD LOW	RECORD PRECIPITATION	RECORD SNOWFALL
January 11	40	–16	1.82″	12.4″
January 12	44	–14	1.35″	13.5″
January 13	45	–19	1.60″	10.0″
January 14	49	–15	0.81″	10.4″
January 15	40	–19	0.66″	17.8″
January 16	44	–21	0.77″	15.0″
January 17	41	–20	0.72″	9.2″
January 18	42	–20	1.13″	**26.5″**
January 19	47	–20	0.48″	8.6″
January 20	43	–21	0.80″	8.6″

Hancock *(continued)*

	RECORD HIGH	RECORD LOW	RECORD PRECIPITATION	RECORD SNOWFALL
January 21	42	–26	0.72″	8.8″
January 22	48	–20	1.02″	12.3″
January 23	44	–20	0.96″	10.7″
January 24	40	–26	0.51″	10.4″
January 25	42	–15	0.90″	12.4″
January 26	47	–22	1.49″	16.5″
January 27	50	–16	0.83″	15.3″
January 28	42	–22	0.60″	12.9″
January 29	43	–15	0.71″	16.3″
January 30	44	–29	0.56″	8.0″
January 31	42	–22	0.84″	8.7″
February 1	42	–22	0.72″	9.0″
February 2	46	–22	0.56″	7.0″
February 3	47	–21	0.48″	7.5″
February 4	44	–23	0.40″	7.0″
February 5	45	–20	0.76″	16.8″
February 6	47	–20	0.96″	12.0″
February 7	42	–20	0.55″	8.9″
February 8	42	–23	1.79″	19.6″
February 9	44	**–30**	0.44″	6.7″
February 10	46	**–30**	0.72″	9.0″
February 11	43	–23	0.72″	9.9″
February 12	40	–20	0.89″	12.0″
February 13	47	–28	0.40″	12.2″
February 14	42	–25	1.27″	13.0″
February 15	44	–21	0.80″	16.2″
February 16	47	–21	0.66″	10.1″
February 17	51	–21	0.54″	9.0″
February 18	51	–20	0.47″	9.0″
February 19	56	–25	0.56″	7.0″
February 20	50	–21	0.95″	10.7″
February 21	47	–17	1.12″	14.0″
February 22	55	–23	0.88″	11.0″
February 23	47	–21	1.09″	14.0″
February 24	48	–17	0.60″	9.5″
February 25	45	–18	0.60″	7.0″
February 26	56	–13	1.27″	9.0″
February 27	50	–19	0.94″	10.0″
February 28	48	–20	0.87″	8.9″
February 29	56	–9	0.17″	4.0″
March 1	46	–14	0.94″	6.0″
March 2	47	–23	0.68″	7.7″
March 3	55	–21	1.39″	16.0″
March 4	53	–12	2.49″	24.9″
March 5	52	–21	0.74″	7.5″
March 6	60	–15	0.56″	8.7″
March 7	56	–21	0.86″	6.4″
March 8	58	–18	0.70″	11.1″
March 9	51	–17	1.21″	14.7″
March 10	55	–11	0.60″	9.0″

	RECORD HIGH	RECORD LOW	RECORD PRECIPITATION	RECORD SNOWFALL
March 11	54	–20	1.06″	7.5″
March 12	58	–12	1.95″	15.5″
March 13	61	–16	1.10″	7.4″
March 14	56	–10	0.96″	12.0″
March 15	64	–9	1.28″	16.0″
March 16	57	–4	0.54″	4.8″
March 17	57	–10	0.56″	8.0″
March 18	56	–13	1.40″	5.5″
March 19	62	–8	1.27″	13.2″
March 20	57	–8	0.64″	7.0″
March 21	60	–9	1.23″	13.5″
March 22	60	–14	1.17″	11.7″
March 23	64	–11	1.06″	7.6″
March 24	64	–11	0.88″	10.8″
March 25	57	–8	1.33″	17.0″
March 26	64	–8	1.01″	7.7″
March 27	69	–8	1.21″	9.0″
March 28	79	–15	0.62″	4.9″
March 29	73	–4	1.15″	10.0″
March 30	72	0	0.65″	7.0″
March 31	65	–6	1.54″	15.6″
April 1	62	1	0.41″	6.0″
April 2	63	4	1.11″	13.5″
April 3	67	–1	0.94″	7.2″
April 4	72	0	0.88″	11.0″
April 5	72	–1	1.00″	9.0″
April 6	71	–4	0.93″	6.6″
April 7	64	7	2.10″	10.0″
April 8	62	7	0.75″	8.4″
April 9	76	4	0.56″	7.0″
April 10	72	–1	1.36″	5.0″
April 11	78	4	1.02″	3.0″
April 12	77	9	0.88″	5.0″
April 13	73	13	2.54″	4.1″
April 14	76	11	0.84″	1.8″
April 15	72	9	1.36″	4.0″
April 16	85	9	0.82″	3.0″
April 17	74	13	1.58″	2.2″
April 18	77	15	1.20″	6.0″
April 19	85	5	1.44″	4.0″
April 20	84	14	0.62″	7.1″
April 21	77	16	1.70″	2.5″
April 22	81	18	1.11″	7.0″
April 23	81	16	0.80″	8.0″
April 24	82	18	1.58″	2.0″
April 25	88	20	1.02″	7.0″
April 26	82	17	2.08″	4.0″
April 27	80	20	1.34″	4.0″
April 28	77	20	1.85″	2.0″
April 29	78	9	0.90″	4.0″
April 30	82	20	1.10″	8.1″

Hancock *(continued)*

	RECORD HIGH	RECORD LOW	RECORD PRECIPITATION	RECORD SNOWFALL
May 1	80	19	0.90″	3.0″
May 2	83	21	1.46″	6.4″
May 3	87	20	1.00″	3.8″
May 4	84	20	1.15″	2.0″
May 5	85	21	1.75″	2.5″
May 6	86	22	1.07″	3.0″
May 7	89	21	1.41″	7.5″
May 8	86	23	1.58″	3.0″
May 9	87	21	1.47″	0.1″
May 10	84	22	0.90″	3.3″
May 11	82	24	1.80″	1.0″
May 12	86	22	1.02″	3.5″
May 13	88	23	1.85″	2.0″
May 14	85	21	1.06″	2.0″
May 15	89	23	1.25″	4.0″
May 16	91	24	1.20″	1.6″
May 17	90	26	**3.58**″	1.0″
May 18	85	24	1.02″	1.0″
May 19	87	21	1.10″	0.2″
May 20	90	26	0.96″	
May 21	89	27	1.25″	0.6″
May 22	94	27	1.96″	0.3″
May 23	87	25	1.45″	
May 24	83	27	1.26″	0.2″
May 25	86	26	0.97″	0.5″
May 26	91	29	2.84″	0.4″
May 27	86	29	1.33″	1.0″
May 28	95	30	0.85″	0.4″
May 29	88	30	1.39″	
May 30	88	30	1.08″	2.0″
May 31	89	28	1.44″	
June 1	87	30	2.20″	
June 2	86	30	1.95″	2.0″
June 3	86	31	1.49″	
June 4	91	32	1.17″	
June 5	92	32	2.50″	
June 6	89	32	1.85″	
June 7	88	33	1.65″	
June 8	90	30	0.84″	
June 9	88	32	2.41″	
June 10	87	32	1.34″	
June 11	88	33	2.31″	
June 12	90	34	2.07″	
June 13	91	34	1.13″	
June 14	96	35	1.35″	
June 15	90	35	1.65″	
June 16	92	36	0.92″	
June 17	92	33	1.35″	
June 18	95	36	0.97″	
June 19	96	37	2.02″	
June 20	95	36	1.15″	

	RECORD HIGH	RECORD LOW	RECORD PRECIPITATION	RECORD SNOWFALL
June 21	91	35	1.02″	
June 22	91	38	1.20″	
June 23	94	36	1.63″	
June 24	91	38	2.19″	
June 25	91	35	2.30″	
June 26	93	38	1.19″	
June 27	91	37	1.65″	
June 28	93	36	1.00″	
June 29	93	34	1.41″	
June 30	99	39	1.38″	
July 1	98	36	1.35″	
July 2	92	40	1.44″	
July 3	93	39	1.36″	
July 4	95	39	3.00″	
July 5	96	37	0.96″	
July 6	99	37	1.53″	
July 7	**102**	42	1.57″	
July 8	99	41	1.47″	
July 9	97	40	1.70″	
July 10	95	32	2.00″	
July 11	92	34	1.46″	
July 12	93	39	1.55″	
July 13	95	41	1.27″	
July 14	97	39	1.28″	
July 15	95	42	1.92″	
July 16	96	42	2.99″	
July 17	95	41	1.57″	
July 18	98	41	2.28″	
July 19	98	39	0.82″	
July 20	93	38	1.73″	
July 21	91	43	2.35″	
July 22	90	42	1.15″	
July 23	96	40	1.40″	
July 24	95	41	1.51″	
July 25	91	45	1.93″	
July 26	91	45	1.42″	
July 27	94	42	1.50″	
July 28	96	42	2.00″	
July 29	101	41	0.74″	
July 30	98	42	1.73″	
July 31	96	41	1.39″	
August 1	92	40	2.25″	
August 2	89	42	1.23″	
August 3	88	39	3.23″	
August 4	90	37	2.00″	
August 5	98	43	1.80″	
August 6	98	45	1.41″	
August 7	92	44	1.78″	
August 8	95	41	2.02″	
August 9	92	40	1.95″	

Hancock *(continued)*

	RECORD HIGH	RECORD LOW	RECORD PRECIPITATION	RECORD SNOWFALL
August 10	91	38	2.50″	
August 11	96	39	1.00″	
August 12	95	39	1.19″	
August 13	88	38	1.50″	
August 14	95	43	1.95″	
August 15	90	39	0.57″	
August 16	90	42	1.35″	
August 17	89	39	1.11″	
August 18	95	35	1.03″	
August 19	97	39	1.71″	
August 20	93	40	2.42″	
August 21	95	40	2.00″	
August 22	90	36	1.85″	
August 23	95	38	1.72″	
August 24	98	39	1.24″	
August 25	92	36	1.52″	
August 26	93	34	1.75″	
August 27	89	39	1.62″	
August 28	92	41	1.50″	
August 29	92	40	1.47″	
August 30	90	37	3.23″	
August 31	93	38	1.81″	
September 1	90	36	1.39″	
September 2	92	33	2.05″	
September 3	92	39	3.06″	
September 4	90	35	**3.58**″	
September 5	88	35	1.94″	
September 6	91	31	2.40″	
September 7	90	31	1.26″	
September 8	89	35	2.70″	
September 9	89	34	1.41″	
September 10	92	32	2.52″	
September 11	95	34	2.05″	
September 12	89	32	0.98″	
September 13	87	31	1.11″	
September 14	90	30	1.40″	
September 15	91	30	1.77″	
September 16	89	30	3.44″	
September 17	88	31	1.63″	
September 18	85	29	2.16″	
September 19	86	28	1.60″	0.2″
September 20	87	28	1.86″	
September 21	88	31	2.09″	
September 22	87	30	1.86″	3.0″
September 23	86	26	1.12″	
September 24	88	26	2.01″	
September 25	82	27	2.40″	
September 26	82	27	1.38″	6.5″
September 27	83	25	1.44″	0.5″
September 28	82	24	2.70″	1.5″
September 29	85	26	1.26″	0.5″

	RECORD HIGH	RECORD LOW	RECORD PRECIPITATION	RECORD SNOWFALL
September 30	79	25	1.13″	0.5″
October 1	83	26	1.09″	0.9″
October 2	86	24	1.94″	0.5″
October 3	79	25	2.75″	1.5″
October 4	78	26	1.39″	
October 5	82	21	2.03″	0.5″
October 6	83	24	1.00″	1.1″
October 7	83	22	0.90″	0.5″
October 8	80	26	0.70″	2.8″
October 9	77	21	1.45″	5.4″
October 10	80	24	0.90″	0.7″
October 11	81	20	1.00″	2.6″
October 12	82	20	1.75″	4.0″
October 13	84	23	0.63″	2.0″
October 14	79	22	2.55″	4.0″
October 15	83	25	1.02″	1.0″
October 16	81	24	0.87″	4.6″
October 17	84	22	1.42″	4.5″
October 18	80	19	1.47″	9.4″
October 19	73	22	1.69″	3.0″
October 20	76	21	1.80″	10.8″
October 21	80	21	0.83″	4.0″
October 22	73	20	1.26″	2.1″
October 23	83	15	2.70″	11.1″
October 24	78	17	1.20″	5.0″
October 25	77	16	1.77″	9.7″
October 26	79	16	1.08″	2.5″
October 27	73	18	1.14″	4.0″
October 28	67	17	1.16″	7.0″
October 29	71	16	1.50″	12.0″
October 30	72	12	0.93″	5.0″
October 31	69	13	0.93″	2.6″
November 1	71	14	1.02″	4.9″
November 2	70	1	1.90″	11.8″
November 3	71	−7	0.82″	5.0″
November 4	68	8	1.13″	4.0″
November 5	64	5	1.85″	7.9″
November 6	65	−4	0.86″	6.9″
November 7	63	−5	1.80″	5.0″
November 8	65	8	1.01″	9.0″
November 9	65	13	0.75″	7.3″
November 10	58	6	1.87″	8.0″
November 11	62	7	1.90″	16.0″
November 12	56	8	1.36″	17.0″
November 13	66	7	1.07″	6.0″
November 14	59	6	0.95″	9.1″
November 15	59	1	0.80″	8.0″
November 16	57	4	0.97″	8.2″
November 17	70	3	1.55″	12.0″
November 18	64	2	1.50″	6.0″

Hancock *(continued)*

	RECORD HIGH	RECORD LOW	RECORD PRECIPITATION	RECORD SNOWFALL
November 19	58	1	1.44″	5.1″
November 20	57	6	1.36″	10.3″
November 21	60	3	0.96″	9.5″
November 22	57	3	0.65″	9.8″
November 23	59	–2	1.28″	12.0″
November 24	55	–2	0.66″	5.0″
November 25	56	0	0.74″	9.0″
November 26	51	–6	1.50″	9.3″
November 27	58	1	0.88″	8.2″
November 28	53	–5	1.13″	12.0″
November 29	53	–3	0.81″	8.2″
November 30	56	0	0.85″	10.9″
December 1	53	–5	1.00″	5.1″
December 2	53	–5	0.80″	8.2″
December 3	54	–4	0.80″	10.0″
December 4	51	–2	1.12″	13.0″
December 5	52	–3	0.82″	11.7″
December 6	52	–3	0.64″	11.5″
December 7	50	–12	1.08″	12.4″
December 8	43	–9	0.88″	13.3″
December 9	48	–7	0.86″	11.7″

	RECORD HIGH	RECORD LOW	RECORD PRECIPITATION	RECORD SNOWFALL
December 10	43	–6	0.76″	7.3″
December 11	51	–5	0.88″	9.5″
December 12	49	–5	0.70″	7.0″
December 13	48	–12	0.66″	7.0″
December 14	51	–10	1.80″	10.7″
December 15	48	–12	0.81″	11.8″
December 16	47	–11	0.88″	11.6″
December 17	43	–6	0.90″	9.0″
December 18	44	–9	0.89″	10.3″
December 19	51	–9	0.49″	10.0″
December 20	46	–10	0.63″	10.2″
December 21	46	–10	0.70″	7.3″
December 22	46	–7	1.53″	15.4″
December 23	43	–10	1.04″	17.8″
December 24	44	–12	0.83″	16.0″
December 25	41	–14	0.43″	6.4″
December 26	44	–13	0.76″	8.5″
December 27	42	–11	0.70″	9.2″
December 28	47	–19	1.19″	14.7″
December 29	40	–15	0.94″	13.3″
December 30	42	–9	0.52″	8.5″
December 31	44	–12	0.92″	9.6″

Harbor Beach (Climate Record Begins 1899)

	RECORD HIGH	RECORD LOW	RECORD PRECIPITATION	RECORD SNOWFALL
January 1	55	–7	1.85″	6.0″
January 2	51	–6	1.21″	7.5″
January 3	56	–8	0.75″	4.5″
January 4	53	–12	0.75″	6.6″
January 5	54	–9	1.14″	5.0″
January 6	50	–12	0.50″	6.0″
January 7	50	–8	0.60″	6.0″
January 8	55	–9	0.82″	8.0″
January 9	55	–6	0.87″	4.5″
January 10	54	–19	0.69″	5.5″
January 11	48	–11	0.75″	5.0″
January 12	44	–13	0.80″	8.0″
January 13	59	–18	0.64″	6.4″
January 14	55	–9	0.95″	6.2″
January 15	53	–13	0.80″	8.0″
January 16	52	–13	0.80″	8.0″
January 17	51	–16	1.00″	8.0″
January 18	52	–14	0.94″	6.0″
January 19	51	–17	1.00″	6.2″
January 20	48	–16	0.86″	8.6″
January 21	53	–14	1.04″	6.0″
January 22	50	–7	0.95″	7.0″
January 23	60	–8	0.60″	6.0″

	RECORD HIGH	RECORD LOW	RECORD PRECIPITATION	RECORD SNOWFALL
January 24	60	–12	0.85″	5.0″
January 25	59	–12	0.45″	6.0″
January 26	62	–10	0.84″	6.2″
January 27	61	–9	1.80″	**18.0″**
January 28	56	–9	0.93″	6.0″
January 29	55	–8	0.70″	7.0″
January 30	49	–19	0.80″	8.0″
January 31	55	–17	1.80″	**18.0″**
February 1	49	–20	1.00″	10.0″
February 2	47	–16	0.80″	8.0″
February 3	42	–20	0.60″	6.0″
February 4	54	–16	1.00″	5.6″
February 5	54	–12	0.95″	7.0″
February 6	51	–11	0.80″	6.0″
February 7	45	–11	1.27″	12.0″
February 8	60	–13	0.80″	6.0″
February 9	52	**–22**	0.47″	3.0″
February 10	50	–21	1.70″	5.5″
February 11	57	–12	1.75″	7.0″
February 12	63	–15	1.51″	6.0″
February 13	48	–14	1.16″	6.0″
February 14	52	–9	1.67″	12.0″

Harbor Beach *(continued)*

	RECORD HIGH	RECORD LOW	RECORD PRECIPITATION	RECORD SNOWFALL
February 15	57	–10	0.65″	6.0″
February 16	59	–18	1.06″	6.0″
February 17	47	–16	1.01″	4.0″
February 18	56	–19	0.62″	5.0″
February 19	55	–12	1.00″	7.3″
February 20	56	–18	0.57″	3.8″
February 21	59	–10	1.80″	**18.0**″
February 22	61	–13	1.30″	10.0″
February 23	68	–4	1.00″	7.0″
February 24	55	–11	0.90″	4.8″
February 25	52	–12	1.16″	11.5″
February 26	52	–11	1.00″	10.0″
February 27	65	–15	0.65″	4.0″
February 28	52	–6	1.00″	6.0″
February 29	48	–5	0.30″	3.0″
March 1	50	–8	1.20″	11.5″
March 2	55	–10	1.45″	7.0″
March 3	63	–8	0.95″	9.5″
March 4	62	–8	1.25″	4.8″
March 5	57	–12	0.95″	5.7″
March 6	63	–11	0.60″	6.0″
March 7	63	–12	0.80″	8.0″
March 8	74	–8	1.20″	6.0″
March 9	80	–5	0.72″	6.0″
March 10	64	–3	0.88″	5.0″
March 11	59	0	0.86″	5.5″
March 12	61	–10	1.10″	4.5″
March 13	72	2	1.01″	4.5″
March 14	72	6	0.76″	5.0″
March 15	68	3	1.55″	15.5″
March 16	69	0	2.41″	3.0″
March 17	74	2	0.71″	6.0″
March 18	64	4	0.75″	5.0″
March 19	74	4	1.46″	5.0″
March 20	73	–1	0.75″	7.5″
March 21	72	3	0.58″	5.7″
March 22	80	5	0.43″	4.3″
March 23	68	2	0.85″	7.0″
March 24	82	–2	0.67″	4.0″
March 25	68	–2	1.00″	8.0″
March 26	74	3	0.63″	6.0″
March 27	77	6	0.93″	5.0″
March 28	79	3	1.55″	3.0″
March 29	84	6	0.65″	1.7″
March 30	76	13	0.84″	2.2″
March 31	75	4	1.14″	3.0″
April 1	80	3	1.50″	4.0″
April 2	81	8	1.34″	8.0″
April 3	78	11	1.60″	16.0″
April 4	76	10	1.12″	4.0″

	RECORD HIGH	RECORD LOW	RECORD PRECIPITATION	RECORD SNOWFALL
April 5	80	15	1.31″	3.0″
April 6	80	6	0.75″	2.0″
April 7	80	14	1.30″	3.5″
April 8	82	15	1.12″	4.0″
April 9	74	17	1.07″	1.5″
April 10	79	16	1.20″	3.4″
April 11	82	20	1.25″	3.4″
April 12	82	18	0.70″	3.6″
April 13	83	17	1.17″	2.0″
April 14	78	19	0.95″	0.5″
April 15	80	20	0.92″	2.0″
April 16	84	21	1.12″	6.0″
April 17	87	19	1.24″	4.0″
April 18	85	19	1.10″	1.0″
April 19	84	21	1.60″	3.2″
April 20	83	21	1.66″	3.6″
April 21	82	20	1.44″	1.2″
April 22	85	20	1.12″	1.5″
April 23	84	23	0.62″	2.5″
April 24	89	24	2.26″	2.5″
April 25	82	24	1.21″	3.0″
April 26	88	22	1.32″	3.2″
April 27	86	22	1.50″	2.0″
April 28	84	22	1.38″	
April 29	82	28	1.27″	2.0″
April 30	83	26	1.30″	3.5″
May 1	83	24	1.07″	2.0″
May 2	88	26	0.81″	4.0″
May 3	87	26	1.05″	
May 4	88	27	0.82″	1.0″
May 5	90	28	1.02″	
May 6	88	28	1.12″	0.5″
May 7	87	27	1.40″	
May 8	88	28	1.36″	6.0″
May 9	88	25	1.96″	1.0″
May 10	86	27	1.32″	
May 11	87	27	2.07″	
May 12	86	25	1.03″	2.8″
May 13	84	24	2.80″	4.3″
May 14	87	32	1.50″	
May 15	88	26	1.40″	
May 16	89	27	1.20″	
May 17	91	28	1.73″	
May 18	87	29	1.61″	
May 19	91	29	1.11″	
May 20	88	29	1.70″	
May 21	92	30	3.00″	0.2″
May 22	88	32	1.47″	0.4″
May 23	88	30	1.48″	
May 24	85	32	1.30″	
May 25	87	34	1.10″	

Harbor Beach *(continued)*

	RECORD HIGH	RECORD LOW	RECORD PRECIPITATION	RECORD SNOWFALL
May 26	90	30	1.85″	
May 27	87	32	0.92″	
May 28	92	31	1.34″	
May 29	90	35	1.41″	
May 30	92	33	0.71″	
May 31	88	37	1.25″	
June 1	100	35	1.62″	
June 2	91	33	1.10″	
June 3	90	35	1.01″	
June 4	90	32	1.20″	
June 5	97	35	1.74″	
June 6	98	37	1.71″	
June 7	95	36	1.13″	
June 8	94	33	1.41″	
June 9	96	35	2.00″	
June 10	94	34	1.74″	
June 11	94	34	2.30″	
June 12	92	38	1.11″	
June 13	92	36	1.60″	
June 14	94	36	1.67″	
June 15	93	38	1.51″	
June 16	96	39	0.74″	
June 17	92	39	4.00″	
June 18	92	38	1.80″	
June 19	93	35	0.80″	
June 20	96	37	1.82″	
June 21	90	39	1.62″	
June 22	94	34	1.05″	
June 23	94	40	1.48″	
June 24	93	39	1.02″	
June 25	96	40	2.10″	
June 26	97	40	1.40″	
June 27	98	43	0.86″	
June 28	98	42	1.35″	
June 29	95	40	1.64″	
June 30	99	40	1.65″	
July 1	101	41	1.77″	
July 2	98	41	1.58″	
July 3	99	38	2.00″	
July 4	97	41	1.51″	
July 5	102	41	1.37″	
July 6	98	40	0.87″	
July 7	95	45	1.40″	
July 8	103	43	0.99″	
July 9	104	43	1.36″	
July 10	**105**	42	1.43″	
July 11	96	43	1.95″	
July 12	95	45	1.33″	
July 13	98	45	1.97″	
July 14	94	44	0.90″	
July 15	100	44	2.71″	
July 16	92	43	0.60″	
July 17	96	42	1.44″	
July 18	98	40	1.51″	
July 19	96	46	1.81″	
July 20	98	42	1.54″	
July 21	95	44	1.63″	
July 22	92	42	0.98″	
July 23	98	44	1.72″	
July 24	102	49	1.95″	
July 25	99	46	1.18″	
July 26	97	46	1.40″	
July 27	95	45	1.01″	
July 28	97	46	0.88″	
July 29	96	46	1.00″	
July 30	100	42	1.02″	
July 31	99	44	1.55″	
August 1	95	44	1.03″	
August 2	96	45	2.27″	
August 3	95	39	1.50″	
August 4	95	47	2.00″	
August 5	100	40	2.27″	
August 6	101	37	1.56″	
August 7	98	43	1.20″	
August 8	97	45	2.51″	
August 9	99	43	1.57″	
August 10	98	44	2.10″	
August 11	96	42	1.74″	
August 12	92	41	2.05″	
August 13	102	41	1.46″	
August 14	98	45	1.60″	
August 15	95	44	2.83″	
August 16	97	43	3.60″	
August 17	97	41	1.30″	
August 18	89	39	1.40″	
August 19	95	38	1.60″	
August 20	94	42	1.28″	
August 21	98	40	2.32″	
August 22	95	42	1.39″	
August 23	96	41	1.72″	
August 24	97	37	1.34″	
August 25	99	38	1.60″	
August 26	96	40	1.50″	
August 27	92	38	1.15″	
August 28	96	35	1.25″	
August 29	94	37	1.50″	
August 30	96	38	1.16″	
August 31	94	40	1.78″	
September 1	98	40	1.10″	
September 2	98	38	1.20″	

Harbor Beach *(continued)*

	RECORD HIGH	RECORD LOW	RECORD PRECIPITATION	RECORD SNOWFALL
September 3	97	41	1.50″	
September 4	93	40	4.16″	
September 5	93	35	0.95″	
September 6	95	38	0.65″	
September 7	94	37	1.12″	
September 8	95	39	2.54″	
September 9	93	35	0.90″	
September 10	94	40	**6.04**″	
September 11	98	34	1.88″	
September 12	92	34	1.39″	
September 13	93	35	2.57″	
September 14	94	30	2.94″	
September 15	96	38	2.22″	
September 16	95	34	2.48″	
September 17	92	32	1.94″	
September 18	93	25	0.98″	
September 19	91	35	1.24″	
September 20	89	33	1.46″	
September 21	93	35	1.10″	
September 22	88	30	0.79″	
September 23	89	28	1.91″	
September 24	85	33	1.00″	
September 25	91	33	0.76″	
September 26	92	31	3.08″	
September 27	91	32	1.37″	
September 28	84	24	1.64″	
September 29	86	21	1.70″	
September 30	80	30	1.11″	
October 1	85	30	1.64″	
October 2	86	30	0.66″	1.5″
October 3	86	30	1.34″	
October 4	87	33	1.31″	
October 5	86	31	1.24″	
October 6	88	27	1.80″	1.4″
October 7	86	24	1.12″	0.1″
October 8	82	29	0.95″	
October 9	88	27	1.02″	
October 10	84	28	1.32″	3.0″
October 11	84	26	0.95″	
October 12	82	24	1.65″	
October 13	83	22	0.97″	1.0″
October 14	81	24	1.31″	
October 15	81	27	1.02″	0.1″
October 16	80	23	1.38″	1.2″
October 17	82	28	1.49″	0.5″
October 18	82	23	1.55″	0.5″
October 19	79	26	0.88″	3.5″
October 20	79	25	0.90″	2.0″
October 21	79	17	0.95″	
October 22	80	25	2.30″	2.0″
October 23	82	24	1.43″	2.0″
October 24	80	21	1.71″	2.6″
October 25	80	23	0.94″	
October 26	76	26	0.66″	3.0″
October 27	78	25	1.06″	3.0″
October 28	72	21	0.95″	0.8″
October 29	71	21	0.53″	1.0″
October 30	76	19	0.86″	
October 31	76	23	0.85″	
November 1	80	21	1.82″	
November 2	76	20	0.87″	1.6″
November 3	74	19	1.34″	4.0″
November 4	74	18	1.58″	1.0″
November 5	75	16	1.80″	3.0″
November 6	71	14	1.42″	2.5″
November 7	70	17	1.51″	6.5″
November 8	74	17	0.94″	4.2″
November 9	73	19	1.35″	5.0″
November 10	73	18	1.12″	1.5″
November 11	72	15	1.07″	2.0″
November 12	68	11	0.88″	2.0″
November 13	67	11	1.20″	3.0″
November 14	69	10	1.42″	3.0″
November 15	65	6	1.36″	5.4″
November 16	70	4	1.18″	0.6″
November 17	68	9	1.14″	4.4″
November 18	68	10	1.26″	3.0″
November 19	71	13	1.29″	6.0″
November 20	65	13	1.02″	4.2″
November 21	67	10	1.75″	4.0″
November 22	68	11	0.80″	5.0″
November 23	69	6	0.83″	2.0″
November 24	65	4	0.73″	3.5″
November 25	56	6	0.98″	4.9″
November 26	65	10	1.04″	8.0″
November 27	62	10	1.70″	6.0″
November 28	67	6	1.20″	3.8″
November 29	60	5	1.05″	4.2″
November 30	62	7	1.43″	8.0″
December 1	65	5	0.80″	3.6″
December 2	60	1	0.85″	4.5″
December 3	64	–1	0.96″	4.3″
December 4	63	0	1.04″	7.0″
December 5	57	8	1.11″	4.5″
December 6	65	5	0.94″	6.0″
December 7	62	5	0.96″	5.0″
December 8	61	1	0.93″	5.5″
December 9	59	1	0.53″	4.3″
December 10	52	3	0.82″	5.6″
December 11	60	0	0.93″	7.0″
December 12	54	2	1.52″	6.5″

Harbor Beach *(continued)*

	RECORD HIGH	RECORD LOW	RECORD PRECIPITATION	RECORD SNOWFALL
December 13	57	0	1.00″	5.0″
December 14	62	–6	0.81″	5.8″
December 15	58	–6	1.28″	4.3″
December 16	53	–4	0.80″	8.0″
December 17	60	–2	0.90″	5.0″
December 18	52	–1	0.64″	8.0″
December 19	57	–6	0.90″	9.0″
December 20	55	–4	0.65″	6.5″
December 21	58	–4	1.18″	14.5″
December 22	60	–5	1.36″	7.0″
December 23	54	–7	0.76″	3.0″
December 24	53	–14	1.14″	7.0″
December 25	64	–6	0.58″	5.0″
December 26	49	–8	0.59″	3.2″
December 27	56	–8	0.84″	5.6″
December 28	59	–7	1.24″	5.0″
December 29	62	–4	0.81″	9.0″
December 30	57	–8	2.69″	6.0″
December 31	61	–8	0.60″	5.0″

Ironwood (Climate Record Begins 1901)

	RECORD HIGH	RECORD LOW	RECORD PRECIPITATION	RECORD SNOWFALL
January 1	46	–25	0.99″	8.0″
January 2	42	–29	0.63″	7.0″
January 3	39	–23	0.58″	7.0″
January 4	41	–23	1.40″	20.0″
January 5	42	–30	0.93″	18.0″
January 6	43	–36	0.62″	17.0″
January 7	45	–37	0.93″	15.0″
January 8	47	–31	0.80″	9.0″
January 9	49	–40	0.50″	12.0″
January 10	48	–35	0.40″	9.8″
January 11	41	–35	0.70″	15.5″
January 12	45	–33	0.90″	9.0″
January 13	46	–25	1.42″	8.8″
January 14	49	–30	0.84″	12.5″
January 15	41	–37	0.82″	12.5″
January 16	45	–35	0.61″	9.0″
January 17	45	**–41**	0.89″	14.5″
January 18	45	–32	0.91″	9.0″
January 19	47	–31	1.46″	18.0″
January 20	48	–31	1.00″	10.0″
January 21	46	–31	0.55″	14.5″
January 22	46	–26	0.97″	12.5″
January 23	47	–34	0.95″	14.3″
January 24	46	–35	0.54″	15.5″
January 25	51	–30	0.70″	12.5″
January 26	55	–33	0.81″	14.0″
January 27	48	–33	0.68″	14.0″
January 28	50	–35	0.58″	12.3″
January 29	45	–30	0.51″	7.0″
January 30	43	–34	1.30″	13.0″
January 31	45	–38	0.81″	6.0″
February 1	44	–36	0.81″	7.0″
February 2	43	–37	0.63″	13.0″
February 3	49	–37	0.74″	11.0″
February 4	49	–37	0.80″	15.0″
February 5	55	–35	1.00″	15.0″
February 6	53	–27	1.50″	15.0″
February 7	51	–28	0.79″	10.5″
February 8	49	–35	1.10″	7.0″
February 9	52	–34	0.43″	6.0″
February 10	47	–34	1.60″	16.0″
February 11	53	–33	0.61″	8.0″
February 12	48	**–41**	1.12″	11.0″
February 13	49	–36	0.61″	9.9″
February 14	47	–31	0.53″	5.0″
February 15	46	–28	1.20″	11.5″
February 16	50	–31	0.55″	12.6″
February 17	55	–36	0.45″	8.0″
February 18	58	–25	0.88″	16.4″
February 19	60	–38	0.69″	8.0″
February 20	56	–31	0.58″	8.0″
February 21	51	–22	1.09″	19.0″
February 22	56	–23	1.23″	11.0″
February 23	60	–24	3.40″	21.0″
February 24	52	–28	1.40″	14.0″
February 25	62	–33	1.01″	12.0″
February 26	55	–29	1.08″	11.0″
February 27	55	–20	0.59″	8.0″
February 28	51	–24	1.10″	12.5″
February 29	50	–12	0.18″	4.0″
March 1	59	–33	1.17″	6.0″
March 2	56	–27	0.65″	15.5″
March 3	58	–28	1.20″	12.0″
March 4	55	–18	1.52″	14.0″
March 5	53	–22	0.82″	14.0″
March 6	60	–25	0.90″	17.0″
March 7	66	–34	1.63″	9.0″
March 8	71	–30	2.09″	17.0″

Ironwood *(continued)*

	RECORD HIGH	RECORD LOW	RECORD PRECIPITATION	RECORD SNOWFALL
March 9	69	–25	1.64″	12.0″
March 10	56	–21	0.60″	9.2″
March 11	58	–20	0.90″	9.0″
March 12	59	–23	0.95″	7.5″
March 13	58	–17	0.70″	11.0″
March 14	63	–16	1.08″	18.0″
March 15	62	–17	0.78″	14.3″
March 16	67	–14	1.09″	7.0″
March 17	67	–17	1.40″	18.0″
March 18	63	–23	1.19″	13.8″
March 19	64	–19	0.82″	12.0″
March 20	64	–24	0.73″	11.0″
March 21	65	–12	1.08″	10.0″
March 22	66	–14	1.20″	13.0″
March 23	72	–19	1.06″	10.0″
March 24	70	–27	1.21″	9.5″
March 25	70	–14	1.40″	18.0″
March 26	66	–23	0.74″	8.5″
March 27	76	–18	1.25″	10.0″
March 28	79	–11	1.54″	8.5″
March 29	78	–8	0.91″	18.1″
March 30	72	–8	1.78″	15.5″
March 31	67	–17	0.65″	5.0″
April 1	74	–2	1.03″	12.0″
April 2	65	0	0.65″	6.0″
April 3	75	–1	0.89″	6.0″
April 4	76	–5	1.24″	13.2″
April 5	75	–4	0.73″	11.3″
April 6	77	–8	1.00″	6.0″
April 7	80	–12	0.84″	11.0″
April 8	75	–7	1.34″	10.5″
April 9	75	–1	0.80″	8.0″
April 10	76	–1	1.23″	3.8″
April 11	83	7	1.47″	5.2″
April 12	79	7	0.85″	11.0″
April 13	77	7	1.14″	6.0″
April 14	78	6	1.26″	6.0″
April 15	83	8	1.37″	7.0″
April 16	81	7	0.93″	8.5″
April 17	82	–8	1.13″	4.0″
April 18	78	–5	0.64″	5.0″
April 19	84	8	2.26″	12.0″
April 20	80	11	0.90″	6.0″
April 21	84	18	1.23″	8.0″
April 22	85	17	0.87″	6.0″
April 23	82	12	3.05″	6.0″
April 24	85	16	2.14″	8.0″
April 25	83	18	0.85″	7.0″
April 26	84	16	2.04″	5.4″
April 27	85	18	1.15″	3.0″
April 28	86	17	1.20″	4.0″

	RECORD HIGH	RECORD LOW	RECORD PRECIPITATION	RECORD SNOWFALL
April 29	88	14	1.20″	4.5″
April 30	87	17	1.05″	8.0″
May 1	85	16	1.80″	8.5″
May 2	91	13	1.81″	6.0″
May 3	90	17	1.73″	3.0″
May 4	88	16	1.35″	2.0″
May 5	85	22	1.59″	2.5″
May 6	86	19	1.28″	2.0″
May 7	86	18	1.08″	4.0″
May 8	86	18	1.81″	6.0″
May 9	86	17	1.89″	2.0″
May 10	86	16	0.92″	4.8″
May 11	87	17	0.89″	4.5″
May 12	85	19	4.09″	8.3″
May 13	84	21	1.04″	3.0″
May 14	90	17	1.48″	2.5″
May 15	88	22	1.76″	3.8″
May 16	88	20	1.56″	4.0″
May 17	88	23	1.50″	0.5″
May 18	88	18	1.45″	
May 19	88	22	1.37″	0.5″
May 20	88	24	1.35″	0.2″
May 21	88	28	2.02″	
May 22	90	27	1.19″	1.7″
May 23	89	25	1.60″	
May 24	87	25	1.62″	
May 25	89	25	1.42″	
May 26	90	26	1.12″	
May 27	89	27	1.26″	
May 28	90	29	0.83″	3.7″
May 29	92	24	1.27″	
May 30	92	30	1.63″	
May 31	100	25	1.79″	
June 1	93	30	1.61″	
June 2	90	30	2.21″	
June 3	90	30	1.28″	
June 4	90	25	1.36″	
June 5	90	28	1.61″	
June 6	90	32	1.53″	
June 7	90	31	2.55″	
June 8	89	29	1.05″	
June 9	90	29	1.22″	
June 10	95	26	2.72″	
June 11	93	33	1.94″	
June 12	90	32	2.06″	
June 13	93	32	1.27″	
June 14	91	31	2.27″	
June 15	91	32	1.16″	
June 16	91	27	3.13″	
June 17	92	33	1.75″	

Ironwood *(continued)*

	RECORD HIGH	RECORD LOW	RECORD PRECIPITATION	RECORD SNOWFALL
June 18	92	34	1.62″	
June 19	94	34	2.25″	
June 20	90	30	2.34″	
June 21	92	28	1.82″	
June 22	91	35	1.23″	
June 23	93	30	2.05″	
June 24	96	29	3.81″	
June 25	90	36	2.05″	
June 26	93	37	1.35″	
June 27	97	34	2.11″	
June 28	96	39	1.35″	
June 29	99	36	1.77″	
June 30	97	34	1.20″	
July 1	97	37	4.13″	
July 2	93	31	2.82″	
July 3	94	38	2.61″	
July 4	90	35	2.35″	
July 5	91	35	1.72″	
July 6	92	39	1.38″	
July 7	98	40	1.13″	
July 8	96	36	1.43″	
July 9	98	41	1.45″	
July 10	98	37	1.05″	
July 11	102	41	0.95″	
July 12	103	38	1.52″	
July 13	**104**	40	1.75″	
July 14	95	38	1.07″	
July 15	95	38	1.98″	
July 16	97	42	1.36″	
July 17	92	37	6.70″	
July 18	94	39	0.97″	
July 19	95	38	0.73″	
July 20	96	39	1.44″	
July 21	90	44	**6.72″**	
July 22	95	38	5.00″	
July 23	102	42	1.43″	
July 24	96	42	1.18″	
July 25	98	35	2.93″	
July 26	94	36	1.90″	
July 27	97	38	1.80″	
July 28	96	39	1.48″	
July 29	95	40	3.32″	
July 30	95	38	3.50″	
July 31	97	40	1.68″	
August 1	96	39	2.03″	
August 2	95	37	1.65″	
August 3	93	37	1.53″	
August 4	101	32	1.54″	
August 5	101	37	0.88″	
August 6	98	35	1.25″	
August 7	93	38	2.06″	
August 8	93	40	1.53″	
August 9	93	37	2.05″	
August 10	96	38	2.23″	
August 11	93	33	1.35″	
August 12	93	37	1.00″	
August 13	91	35	1.45″	
August 14	92	31	1.06″	
August 15	97	30	0.95″	
August 16	92	33	4.34″	
August 17	93	37	1.43″	
August 18	92	34	2.00″	
August 19	92	36	0.75″	
August 20	94	36	1.30″	
August 21	89	34	2.03″	
August 22	95	35	2.38″	
August 23	95	31	2.95″	
August 24	98	30	0.87″	
August 25	97	36	1.55″	
August 26	96	34	2.71″	
August 27	91	35	1.43″	
August 28	90	32	1.58″	
August 29	91	31	2.73″	
August 30	90	35	2.61″	
August 31	90	35	2.77″	
September 1	92	32	1.98″	
September 2	92	29	1.72″	
September 3	93	31	2.54″	
September 4	92	32	1.56″	
September 5	91	32	2.32″	
September 6	90	34	2.00″	
September 7	90	32	1.70″	
September 8	99	30	1.45″	
September 9	93	29	1.31″	
September 10	96	32	1.53″	
September 11	92	31	1.30″	
September 12	90	32	1.06″	
September 13	91	26	1.54″	
September 14	93	29	1.61″	
September 15	93	24	2.25″	
September 16	88	28	1.38″	
September 17	89	24	1.97″	
September 18	85	30	1.03″	
September 19	88	27	1.71″	
September 20	87	24	1.26″	
September 21	88	25	1.62″	2.0″
September 22	86	25	0.84″	6.5″
September 23	89	24	1.16″	1.0″
September 24	86	22	1.88″	1.5″
September 25	83	24	1.15″	2.0″
September 26	85	22	2.30″	

Ironwood *(continued)*

	RECORD HIGH	RECORD LOW	RECORD PRECIPITATION	RECORD SNOWFALL
September 27	88	22	1.17″	2.0″
September 28	84	21	1.22″	
September 29	85	21	1.52″	1.0″
September 30	83	26	1.29″	
October 1	80	24	1.52″	4.0″
October 2	86	20	1.85″	3.5″
October 3	86	22	1.16″	1.0″
October 4	86	19	2.01″	
October 5	84	17	3.19″	9.0″
October 6	84	17	2.10″	3.5″
October 7	81	23	1.44″	4.5″
October 8	83	19	1.35″	3.0″
October 9	80	21	1.00″	3.0″
October 10	84	18	0.85″	7.2″
October 11	84	19	1.00″	6.0″
October 12	80	19	1.42″	8.0″
October 13	82	18	0.87″	6.2″
October 14	80	18	1.75″	6.5″
October 15	84	21	2.05″	2.7″
October 16	84	20	1.01″	4.8″
October 17	82	19	1.24″	3.0″
October 18	81	16	1.82″	5.0″
October 19	80	15	0.85″	7.0″
October 20	82	15	2.05″	11.0″
October 21	85	15	1.80″	16.0″
October 22	83	14	0.98″	9.0″
October 23	79	9	1.44″	6.0″
October 24	76	5	1.57″	8.0″
October 25	76	15	1.33″	15.0″
October 26	76	12	1.05″	10.5″
October 27	76	14	1.25″	5.0″
October 28	71	12	0.90″	5.0″
October 29	72	9	1.15″	8.0″
October 30	77	2	1.12″	10.3″
October 31	81	11	1.25″	4.2″
November 1	76	8	5.61″	8.4″
November 2	72	–1	2.26″	20.0″
November 3	69	–4	0.86″	14.0″
November 4	74	2	0.82″	9.5″
November 5	71	7	1.64″	13.0″
November 6	72	0	0.73″	11.0″
November 7	66	–3	1.07″	5.0″
November 8	72	–1	1.78″	12.0″
November 9	72	–7	1.50″	10.0″
November 10	71	5	1.72″	12.0″
November 11	65	–7	0.95″	6.0″
November 12	59	–6	1.17″	10.0″
November 13	59	–8	0.78″	10.0″
November 14	67	–5	0.74″	7.5″
November 15	64	–7	0.96″	16.0″
November 16	63	–2	1.68″	11.0″
November 17	68	–8	1.22″	7.0″
November 18	65	–9	1.63″	9.0″
November 19	66	–12	1.43″	15.9″
November 20	62	–13	0.98″	12.0″
November 21	58	–13	1.05″	12.0″
November 22	65	–15	2.30″	14.0″
November 23	54	–15	1.23″	9.5″
November 24	56	–10	1.70″	12.0″
November 25	55	–10	0.58″	8.0″
November 26	58	–16	1.62″	14.0″
November 27	56	–10	0.86″	9.0″
November 28	65	–15	1.17″	9.0″
November 29	55	–14	0.68″	10.0″
November 30	58	–18	1.54″	15.0″
December 1	55	–20	0.59″	9.0″
December 2	54	–14	0.97″	14.0″
December 3	58	–20	0.72″	10.0″
December 4	59	–16	0.70″	11.0″
December 5	50	–11	1.13″	11.0″
December 6	55	–16	0.87″	9.0″
December 7	51	–22	1.25″	11.0″
December 8	47	–22	1.49″	12.0″
December 9	54	–24	1.31″	9.0″
December 10	50	–22	0.70″	22.1″
December 11	51	–26	0.68″	10.5″
December 12	50	–20	0.70″	8.0″
December 13	52	–25	0.93″	12.0″
December 14	43	–21	0.61″	14.8″
December 15	51	–26	0.50″	10.0″
December 16	47	–20	2.09″	**24.0″**
December 17	49	–19	0.61″	10.0″
December 18	49	–27	0.63″	10.0″
December 19	51	–31	0.95″	14.5″
December 20	46	–36	1.42″	17.5″
December 21	48	–28	0.55″	5.2″
December 22	42	–24	0.68″	12.0″
December 23	42	–26	0.76″	21.0″
December 24	46	–23	0.84″	15.0″
December 25	53	–28	0.97″	12.0″
December 26	52	–22	0.40″	12.0″
December 27	45	–24	1.20″	19.5″
December 28	42	–25	1.10″	8.0″
December 29	45	–29	0.70″	11.0″
December 30	43	–35	0.63″	6.0″
December 31	47	–27	1.54″	14.0″

Jackson (Climate Record Begins 1897)

	RECORD HIGH	RECORD LOW	RECORD PRECIPITATION	RECORD SNOWFALL
January 1	59	–9	1.16″	10.0″
January 2	57	–10	0.47″	3.0″
January 3	58	–10	0.78″	14.0″
January 4	60	–13	1.30″	3.0″
January 5	56	–10	0.68″	2.5″
January 6	66	–12	1.25″	4.0″
January 7	71	–12	0.65″	5.0″
January 8	60	–12	1.55″	2.0″
January 9	58	–10	0.82″	9.6″
January 10	57	–13	0.80″	9.0″
January 11	55	–15	0.40″	6.2″
January 12	59	–19	1.80″	8.5″
January 13	61	–17	1.09″	11.0″
January 14	56	–10	0.76″	3.8″
January 15	61	–17	0.50″	4.2″
January 16	56	–18	0.52″	6.0″
January 17	63	–16	0.56″	3.0″
January 18	60	–20	1.52″	5.0″
January 19	62	–20	0.92″	5.0″
January 20	63	–18	1.10″	2.0″
January 21	59	–18	1.00″	4.5″
January 22	59	–13	1.07″	5.2″
January 23	60	–17	0.66″	4.8″
January 24	60	–17	0.60″	4.0″
January 25	67	–20	0.42″	4.7″
January 26	65	–15	1.43″	14.1″
January 27	64	–12	0.90″	6.0″
January 28	57	–11	0.50″	2.1″
January 29	57	–12	0.81″	5.0″
January 30	53	–12	0.85″	7.0″
January 31	60	–9	1.25″	11.5″
February 1	54	–10	1.63″	6.0″
February 2	50	–12	0.42″	4.0″
February 3	57	–14	0.80″	8.0″
February 4	57	–12	0.72″	6.0″
February 5	52	–20	0.85″	5.0″
February 6	57	–7	0.73″	6.0″
February 7	54	–17	0.60″	5.0″
February 8	64	–17	0.60″	6.0″
February 9	58	–18	1.27″	6.0″
February 10	61	**–21**	0.89	4.6″
February 11	69	–12	0.77″	5.7″
February 12	63	–19	0.71″	3.0″
February 13	63	–17	0.83″	5.0″
February 14	61	–14	1.40″	5.0″
February 15	64	–11	0.70″	3.5″
February 16	60	–12	2.27″	4.0″
February 17	59	–17	0.88″	5.0″
February 18	59	–14	0.97″	5.5″
February 19	62	–12	1.98″	10.0″
February 20	62	–13	0.95″	6.0″
February 21	64	–10	0.80″	9.0″
February 22	64	–9	0.65″	4.7″
February 23	65	–5	0.85″	3.0″
February 24	61	–7	0.92″	3.2″
February 25	65	–11	0.98″	7.0″
February 26	63	–11	0.83″	4.0″
February 27	67	–14	1.22″	3.0″
February 28	57	–6	1.00″	10.0″
February 29	58	–5	0.07″	0.6″
March 1	64	–4	0.60″	4.0″
March 2	61	–6	0.97″	4.5″
March 3	71	–7	0.99″	7.0″
March 4	72	–4	1.98″	4.9″
March 5	68	–4	0.80″	8.0″
March 6	69	–1	1.10″	8.0″
March 7	76	–5	0.73″	4.0″
March 8	79	–5	0.97″	6.0″
March 9	66	–7	1.03″	10.0″
March 10	69	–4	0.85″	8.5″
March 11	72	0	0.70″	7.4″
March 12	75	–1	1.11″	3.4″
March 13	72	3	1.22″	2.0″
March 14	76	1	1.33″	8.0″
March 15	76	6	1.60″	10.0″
March 16	73	1	1.00″	3.0″
March 17	72	–3	0.83″	**16.0**″
March 18	76	4	0.76″	5.7″
March 19	77	0	1.68″	4.0″
March 20	77	8	0.66″	4.0″
March 21	77	8	0.99″	5.0″
March 22	83	7	0.78″	4.1″
March 23	79	0	0.71″	10.0″
March 24	80	–4	1.02″	2.6″
March 25	78	3	1.39″	5.5″
March 26	78	6	1.37″	5.5″
March 27	80	–4	1.21″	7.5″
March 28	78	–6	0.88″	3.0″
March 29	77	10	1.25″	8.0″
March 30	79	7	1.45″	5.0″
March 31	79	4	1.82″	4.0″
April 1	78	9	1.45″	5.5″
April 2	80	17	1.28″	2.8″
April 3	77	11	1.11″	4.0″
April 4	77	14	1.75″	4.0″
April 5	80	13	2.70″	7.2″
April 6	88	15	1.55″	4.0″
April 7	82	3	1.65″	4.0″
April 8	83	9	0.76″	0.8″

Jackson *(continued)*

	RECORD HIGH	RECORD LOW	RECORD PRECIPITATION	RECORD SNOWFALL
April 9	79	15	0.64″	2.5″
April 10	83	16	1.20″	4.0″
April 11	84	21	1.94″	2.3″
April 12	85	16	0.76″	2.2″
April 13	85	17	1.43″	0.5″
April 14	82	20	1.08″	1.0″
April 15	83	18	1.65″	1.5″
April 16	84	18	1.43″	3.0″
April 17	83	22	1.88″	3.0″
April 18	88	21	1.26″	1.0″
April 19	83	18	1.16″	0.3″
April 20	82	17	2.46″	
April 21	84	22	1.40″	2.0″
April 22	88	22	1.15″	
April 23	87	21	0.96″	4.0″
April 24	87	24	1.17″	1.6″
April 25	87	23	0.93″	0.3″
April 26	89	24	0.91″	
April 27	84	24	0.94″	2.3″
April 28	82	24	1.30″	
April 29	88	26	2.27″	1.0″
April 30	87	26	1.70″	
May 1	87	23	1.36″	0.5″
May 2	87	27	1.03″	1.0″
May 3	89	28	0.85″	0.5″
May 4	87	24	1.10″	
May 5	89	29	0.73″	
May 6	89	29	1.79″	
May 7	88	25	0.84″	
May 8	90	27	1.89″	
May 9	88	25	1.31″	4.0″
May 10	89	21	1.58″	1.0″
May 11	91	26	2.60″	
May 12	87	28	2.17″	
May 13	88	27	0.95″	
May 14	89	30	1.60″	
May 15	91	30	2.45″	
May 16	91	30	1.25″	
May 17	92	32	1.24″	
May 18	92	30	3.06″	
May 19	95	28	1.55″	
May 20	92	31	0.86″	
May 21	93	30	2.04″	
May 22	90	31	1.59″	
May 23	93	29	0.94″	
May 24	91	31	0.95″	
May 25	92	30	1.39″	
May 26	95	34	1.39″	
May 27	99	30	1.85″	
May 28	94	31	1.53″	
May 29	93	30	1.56″	

	RECORD HIGH	RECORD LOW	RECORD PRECIPITATION	RECORD SNOWFALL
May 30	95	31	1.75″	
May 31	97	32	1.73″	
June 1	101	34	1.77″	
June 2	100	34	1.72″	
June 3	95	33	0.81″	
June 4	96	34	1.25″	
June 5	98	34	2.11″	
June 6	98	36	2.15″	
June 7	97	38	1.49″	
June 8	96	37	2.00″	
June 9	94	35	1.42″	
June 10	95	34	1.51″	
June 11	96	36	1.15″	
June 12	95	39	1.38″	
June 13	94	42	1.60″	
June 14	96	37	1.16″	
June 15	95	35	1.20″	
June 16	97	39	2.58″	
June 17	95	40	1.98″	
June 18	95	41	1.44″	
June 19	95	44	1.52″	
June 20	98	38	1.61″	
June 21	97	36	**5.31**″	
June 22	97	39	1.76″	
June 23	99	36	3.00″	
June 24	98	38	2.64″	
June 25	100	37	2.60″	
June 26	96	44	1.59″	
June 27	101	42	1.44″	
June 28	102	40	1.44″	
June 29	99	42	1.92″	
June 30	100	41	1.50″	
July 1	99	40	1.38″	
July 2	100	37	1.69″	
July 3	102	42	1.98″	
July 4	101	44	1.57″	
July 5	103	43	2.78″	
July 6	99	42	1.07″	
July 7	99	45	3.38″	
July 8	103	43	1.59″	
July 9	103	47	1.42″	
July 10	101	44	1.03″	
July 11	100	41	2.84″	
July 12	104	47	0.95″	
July 13	103	43	1.63″	
July 14	**105**	43	1.50″	
July 15	103	42	2.05″	
July 16	97	44	1.20″	
July 17	99	48	1.98″	
July 18	97	45	2.45″	

Jackson *(continued)*

	RECORD HIGH	RECORD LOW	RECORD PRECIPITATION	RECORD SNOWFALL
July 19	100	43	1.57″	
July 20	100	42	2.10″	
July 21	100	44	2.25″	
July 22	99	45	1.40″	
July 23	98	45	1.21″	
July 24	104	45	1.61″	
July 25	102	45	1.60″	
July 26	99	46	1.33″	
July 27	101	47	2.18″	
July 28	99	47	1.83″	
July 29	104	44	2.40″	
July 30	102	45	1.12″	
July 31	100	42	1.79″	
August 1	96	46	2.30″	
August 2	102	41	2.09″	
August 3	99	42	1.80″	
August 4	97	46	1.13″	
August 5	99	43	1.37″	
August 6	103	42	1.89″	
August 7	101	45	1.14″	
August 8	98	43	2.40″	
August 9	98	41	2.60″	
August 10	98	43	1.04″	
August 11	97	44	1.74″	
August 12	96	41	1.25″	
August 13	98	44	2.25″	
August 14	99	41	1.90″	
August 15	99	42	1.26″	
August 16	101	40	1.39″	
August 17	95	42	2.34″	
August 18	96	41	1.48″	
August 19	99	40	1.39″	
August 20	101	42	2.48″	
August 21	101	37	1.88″	
August 22	101	38	1.50″	
August 23	98	39	1.91″	
August 24	96	41	0.99″	
August 25	96	40	1.37″	
August 26	95	45	2.26″	
August 27	98	41	1.61″	
August 28	94	38	0.74″	
August 29	95	37	2.03″	
August 30	97	37	1.99″	
August 31	96	38	1.20″	
September 1	96	41	1.61″	
September 2	101	37	1.76″	
September 3	97	34	1.94″	
September 4	94	38	3.17″	
September 5	97	40	1.52″	
September 6	95	35	1.25″	
September 7	96	36	1.05″	
September 8	100	36	1.95″	
September 9	96	38	1.09″	
September 10	96	36	0.84″	
September 11	97	34	1.12″	
September 12	95	35	2.02″	
September 13	93	33	2.80″	
September 14	98	33	2.23″	
September 15	100	36	1.66″	
September 16	97	34	1.11″	
September 17	92	30	2.14″	
September 18	92	35	1.99″	
September 19	93	29	1.00″	
September 20	93	30	2.74″	
September 21	94	30	1.53″	
September 22	92	31	2.56″	
September 23	90	30	2.70″	
September 24	92	28	1.90″	
September 25	90	31	2.25″	
September 26	91	30	1.92″	
September 27	87	30	2.21″	
September 28	92	27	1.60″	
September 29	95	27	1.20″	
September 30	88	27	1.98″	
October 1	90	23	1.30″	
October 2	91	26	0.83″	
October 3	89	26	1.74″	
October 4	90	28	1.36″	
October 5	88	24	1.43″	
October 6	87	25	1.44″	
October 7	87	22	1.36″	
October 8	88	25	0.97″	
October 9	86	23	1.05″	
October 10	85	24	1.17″	
October 11	84	23	1.95″	
October 12	85	26	3.80″	
October 13	84	25	0.89″	
October 14	87	24	1.35″	
October 15	86	24	1.44″	
October 16	87	24	1.98″	1.0″
October 17	87	21	1.53″	
October 18	85	18	1.93″	
October 19	83	19	1.97″	3.4″
October 20	84	20	0.81″	0.7″
October 21	83	19	0.86″	0.5″
October 22	85	22	1.60″	
October 23	82	21	2.58″	2.0″
October 24	81	22	0.84″	0.1″
October 25	80	22	0.92″	
October 26	80	23	1.10″	0.1″
October 27	81	16	1.82″	2.0″

Jackson *(continued)*

	RECORD HIGH	RECORD LOW	RECORD PRECIPITATION	RECORD SNOWFALL
October 28	80	20	0.77″	3.0″
October 29	77	16	0.51″	0.5″
October 30	77	18	1.65″	1.0″
October 31	79	16	1.10″	
November 1	79	19	1.58″	
November 2	77	18	1.41″	4.0″
November 3	75	14	0.70″	5.0″
November 4	75	13	0.92″	1.1″
November 5	77	10	1.64″	1.0″
November 6	76	17	1.45″	6.3″
November 7	72	15	2.67″	0.9″
November 8	74	12	0.80″	3.0″
November 9	73	13	1.05″	5.0″
November 10	69	18	1.41″	4.0″
November 11	72	15	0.74″	1.0″
November 12	72	14	2.20″	3.4″
November 13	70	8	1.38″	2.1″
November 14	68	11	1.01″	2.0″
November 15	68	7	0.86″	8.0″
November 16	68	5	1.43″	6.0″
November 17	69	8	0.74″	2.0″
November 18	69	8	1.05″	1.1″
November 19	69	9	1.38″	4.3″
November 20	72	11	1.17″	3.0″
November 21	69	11	1.08″	2.0″
November 22	67	11	0.91″	5.0″
November 23	69	4	0.68″	3.0″
November 24	65	–5	0.72″	5.6″
November 25	64	–2	1.21″	3.3″
November 26	63	–5	1.01″	1.7″
November 27	67	7	1.30″	5.0″
November 28	69	4	1.29″	2.0″
November 29	65	2	1.54″	4.4″
November 30	63	2	1.20″	3.5″

	RECORD HIGH	RECORD LOW	RECORD PRECIPITATION	RECORD SNOWFALL
December 1	65	–10	0.74″	8.7″
December 2	67	–10	0.89″	3.1″
December 3	67	–10	0.79″	2.5″
December 4	65	–2	0.56″	3.0″
December 5	69	5	0.90″	6.5″
December 6	68	–1	1.20″	4.0″
December 7	60	–3	1.37″	4.0″
December 8	64	–1	0.57″	5.0″
December 9	58	–6	0.53″	3.0″
December 10	63	–3	0.94″	10.8″
December 11	62	–9	0.68″	4.0″
December 12	62	–9	1.14″	6.0″
December 13	57	–3	1.00″	4.0″
December 14	66	–12	0.87″	7.0″
December 15	59	–13	0.91″	5.0″
December 16	63	–7	1.25″	3.0″
December 17	55	–6	1.10″	5.2″
December 18	58	–3	0.84″	7.0″
December 19	55	–6	0.65″	7.7″
December 20	58	–6	0.69″	5.6″
December 21	61	–7	0.78″	6.0″
December 22	58	–14	0.85″	6.0″
December 23	55	–12	1.16″	7.2″
December 24	58	–14	2.03″	3.1″
December 25	64	–12	0.60″	6.0″
December 26	52	–11	0.84″	5.0″
December 27	61	–8	1.02″	4.5″
December 28	64	–14	0.58″	8.0″
December 29	62	–5	2.03″	4.6″
December 30	58	–8	1.33″	2.0″
December 31	59	–12	1.31″	7.0″

Munising (Climate Record Begins 1911)

	RECORD HIGH	RECORD LOW	RECORD PRECIPITATION	RECORD SNOWFALL
January 1	41	–19	0.98″	10.0″
January 2	45	–18	0.52″	7.5″
January 3	42	–15	1.40″	11.0″
January 4	41	–16	0.97″	10.0″
January 5	42	–16	0.88″	9.0″
January 6	44	–14	0.48″	5.1″
January 7	48	–17	0.66″	9.0″
January 8	44	–17	0.76″	10.0″
January 9	46	–15	0.60″	8.0″
January 10	46	–17	0.49″	6.5″
January 11	45	–15	0.53″	12.5″
January 12	58	–15	0.57″	6.5″
January 13	43	–20	0.67″	6.5″
January 14	42	–18	1.21″	10.0″
January 15	40	–20	0.60″	7.0″
January 16	40	–18	0.57″	11.0″
January 17	47	–21	0.55″	9.0″
January 18	47	–25	0.70″	8.5″
January 19	46	–27	1.07″	8.0″
January 20	41	–21	0.54″	10.7″

Munising *(continued)*

	RECORD HIGH	RECORD LOW	RECORD PRECIPITATION	RECORD SNOWFALL
January 21	42	−21	0.71″	8.0″
January 22	45	−18	0.87″	9.3″
January 23	45	−23	0.50″	8.5″
January 24	45	−27	0.89″	8.5″
January 25	44	−23	1.75″	17.5″
January 26	50	−22	1.15″	15.0″
January 27	48	−17	0.72″	5.0″
January 28	47	−18	0.91″	10.1″
January 29	43	−15	0.79″	9.0″
January 30	42	−26	0.68″	8.0″
January 31	42	−21	0.51″	8.2″
February 1	40	−29	0.82″	6.0″
February 2	49	−30	0.52″	7.0″
February 3	52	−21	0.40″	6.6″
February 4	54	−24	0.57″	13.5″
February 5	45	−28	0.42″	8.0″
February 6	50	−21	0.60″	6.0″
February 7	49	−27	0.68″	8.0″
February 8	52	−24	0.36″	5.0″
February 9	45	−26	0.76″	7.0″
February 10	44	−25	0.54″	6.0″
February 11	47	−15	0.74″	7.5″
February 12	46	−24	1.16″	10.0″
February 13	41	−17	0.84″	10.5″
February 14	41	−20	0.79″	9.8″
February 15	48	−15	0.83″	12.7″
February 16	52	−27	0.55″	8.0″
February 17	57	−27	0.73″	9.0″
February 18	51	−22	1.40″	6.0″
February 19	53	−31	0.45″	8.0″
February 20	47	−31	1.06″	10.6″
February 21	48	−23	2.67″	8.0″
February 22	55	−22	1.17″	11.7″
February 23	49	−16	1.25″	12.5″
February 24	44	−17	0.70″	5.6″
February 25	53	**−33**	0.75″	8.0″
February 26	51	−24	0.75″	7.5″
February 27	51	−15	0.75″	7.0″
February 28	49	−18	0.76″	12.0″
February 29	52	−7	0.10″	3.0″
March 1	48	−18	0.60″	6.0″
March 2	51	−22	0.45″	7.0″
March 3	51	−26	0.58″	6.4″
March 4	50	−18	1.30″	10.0″
March 5	47	−22	0.42″	5.0″
March 6	52	−20	2.23″	13.5″
March 7	53	−23	0.58″	6.5″
March 8	59	−26	0.70″	10.0″
March 9	70	−19	0.70″	7.5″
March 10	57	−16	0.90″	6.5″

	RECORD HIGH	RECORD LOW	RECORD PRECIPITATION	RECORD SNOWFALL
March 11	56	−25	1.25″	9.0″
March 12	54	−11	0.54″	6.0″
March 13	60	−10	0.87″	8.7″
March 14	66	−11	1.19″	10.0″
March 15	56	−15	2.75″	**20.0″**
March 16	57	−8	0.70″	6.8″
March 17	55	−18	0.83″	6.0″
March 18	62	−15	0.59″	7.0″
March 19	66	−16	1.86″	17.0″
March 20	65	−14	1.51″	6.0″
March 21	57	−6	1.05″	9.0″
March 22	55	−16	1.25″	10.5″
March 23	58	−17	0.62″	6.0″
March 24	62	−6	1.64″	10.1″
March 25	66	−10	1.31″	12.5″
March 26	66	−22	0.74″	7.0″
March 27	71	−8	0.84″	4.5″
March 28	80	−9	1.02″	6.3″
March 29	67	−13	0.94″	8.0″
March 30	64	−5	0.86″	6.0″
March 31	66	−17	0.43″	4.1″
April 1	74	−15	1.14″	9.0″
April 2	63	0	0.76″	7.0″
April 3	65	5	0.85″	4.0″
April 4	69	−8	0.98″	10.0″
April 5	76	0	1.28″	13.3″
April 6	75	−4	0.79″	10.1″
April 7	75	0	1.33″	9.0″
April 8	69	2	0.91″	6.8″
April 9	75	2	1.15″	5.0″
April 10	79	3	2.04″	10.0″
April 11	75	8	1.35″	6.0″
April 12	79	4	0.79″	5.5″
April 13	75	2	2.26″	7.0″
April 14	78	0	0.72″	5.0″
April 15	80	9	0.75″	6.0″
April 16	76	16	1.22″	5.5″
April 17	82	10	1.22″	3.0″
April 18	79	12	0.54″	2.0″
April 19	73	10	1.10″	7.0″
April 20	81	−3	0.97″	5.2″
April 21	78	17	0.80″	2.0″
April 22	82	17	1.10″	11.0″
April 23	80	18	1.35″	5.0″
April 24	85	17	0.67″	2.0″
April 25	83	18	1.52″	3.5″
April 26	89	19	1.02″	8.0″
April 27	86	18	1.32″	2.0″
April 28	81	17	2.45″	7.0″
April 29	83	20	1.14″	1.5″
April 30	82	15	1.04″	4.0″

Munising *(continued)*

	RECORD HIGH	RECORD LOW	RECORD PRECIPITATION	RECORD SNOWFALL
May 1	86	21	1.16″	2.0″
May 2	79	21	1.30″	1.5″
May 3	89	20	1.10″	1.0″
May 4	87	10	0.97″	3.0″
May 5	93	17	3.01″	1.5″
May 6	87	20	1.82″	3.6″
May 7	88	21	1.99″	7.0″
May 8	82	20	1.03″	
May 9	86	20	1.67″	1.0″
May 10	86	20	0.81″	6.0″
May 11	87	20	0.69″	0.4″
May 12	82	20	1.76″	1.0″
May 13	84	16	1.17″	0.1″
May 14	89	20	2.18″	0.5″
May 15	90	20	1.42″	0.5″
May 16	92	23	1.06″	2.0″
May 17	90	22	0.61″	
May 18	84	23	1.12″	
May 19	87	23	1.81″	
May 20	89	21	1.05″	1.3″
May 21	92	22	2.22″	
May 22	89	22	1.17″	
May 23	89	21	1.08″	10.0″
May 24	87	22	1.32″	1.0″
May 25	87	24	0.98″	
May 26	87	24	1.97″	
May 27	92	25	1.16″	
May 28	95	25	1.10″	
May 29	91	26	2.00″	6.0″
May 30	92	27	2.31″	
May 31	93	27	**3.51**″	
June 1	94	22	1.41″	
June 2	93	26	2.68″	
June 3	90	25	1.20″	
June 4	90	28	0.80″	
June 5	91	29	0.90″	
June 6	93	29	1.04″	
June 7	94	29	1.25″	
June 8	94	25	1.34″	
June 9	93	27	0.83″	
June 10	91	29	2.55″	
June 11	94	30	2.25″	
June 12	95	29	2.00″	
June 13	96	21	1.50″	
June 14	86	30	1.00″	
June 15	89	29	1.67″	
June 16	90	30	2.16″	
June 17	94	28	1.86″	
June 18	96	30	0.58″	
June 19	97	28	1.27″	
June 20	88	29	1.48″	

	RECORD HIGH	RECORD LOW	RECORD PRECIPITATION	RECORD SNOWFALL
June 21	88	29	1.29″	
June 22	92	32	1.93″	
June 23	95	26	1.52″	
June 24	91	31	1.40″	
June 25	91	34	0.95″	
June 26	95	30	1.28″	
June 27	90	32	1.76″	
June 28	93	30	1.53″	
June 29	96	32	1.64″	
June 30	96	30	1.46″	
July 1	99	31	1.72″	
July 2	101	33	1.30″	
July 3	93	33	1.43″	
July 4	95	32	1.56″	
July 5	95	34	1.34″	
July 6	99	35	1.27″	
July 7	**103**	35	1.88″	
July 8	**103**	32	1.62″	
July 9	**103**	33	2.10″	
July 10	98	35	1.49″	
July 11	95	35	3.12″	
July 12	94	33	2.26″	
July 13	100	37	1.05″	
July 14	100	34	1.06″	
July 15	99	34	2.75″	
July 16	92	36	1.12″	
July 17	93	35	2.54″	
July 18	95	35	1.36″	
July 19	96	34	1.01″	
July 20	96	34	2.26″	
July 21	95	34	1.83″	
July 22	92	33	1.32″	
July 23	95	35	2.40″	
July 24	96	38	2.14″	
July 25	94	37	2.02″	
July 26	96	35	1.14″	
July 27	93	36	1.77″	
July 28	98	37	1.45″	
July 29	101	33	2.12″	
July 30	99	32	2.22″	
July 31	93	38	1.88″	
August 1	96	33	1.58″	
August 2	97	28	2.13″	
August 3	97	34	1.80″	
August 4	95	34	1.17″	
August 5	99	35	1.30″	
August 6	**103**	38	1.43″	
August 7	92	34	1.20″	
August 8	97	34	1.18″	
August 9	96	33	2.05″	

Munising *(continued)*

	RECORD HIGH	RECORD LOW	RECORD PRECIPITATION	RECORD SNOWFALL
August 10	93	34	1.04″	
August 11	96	35	0.75″	
August 12	98	33	0.80″	
August 13	96	36	1.75″	
August 14	95	34	1.45″	
August 15	95	32	2.50″	
August 16	92	26	1.78″	
August 17	91	33	1.89″	
August 18	94	34	1.40″	
August 19	97	37	0.75″	
August 20	95	34	3.16″	
August 21	97	32	1.69″	
August 22	89	34	1.92″	
August 23	98	33	1.45″	
August 24	99	32	1.00″	
August 25	97	24	1.35″	
August 26	96	32	2.05″	
August 27	95	32	2.00″	
August 28	96	33	1.53″	
August 29	98	28	1.75″	
August 30	95	34	1.83″	
August 31	96	34	1.50″	
September 1	98	35	1.04″	
September 2	91	30	1.42″	
September 3	91	30	1.78″	
September 4	90	31	2.45″	
September 5	94	31	1.40″	
September 6	91	32	1.97″	
September 7	93	31	1.29″	
September 8	94	33	1.64″	
September 9	99	32	1.00″	
September 10	91	30	1.21″	
September 11	97	27	2.34″	
September 12	93	29	1.16″	
September 13	91	30	1.53″	
September 14	94	29	2.06″	
September 15	95	34	2.10″	
September 16	90	29	1.41″	
September 17	85	29	1.36″	
September 18	90	32	1.46″	
September 19	85	27	1.43″	
September 20	86	29	1.50″	
September 21	85	27	1.46″	
September 22	81	29	2.13″	0.1″
September 23	83	27	1.03″	
September 24	81	28	1.69″	
September 25	84	23	0.82″	2.0″
September 26	81	24	1.80″	
September 27	84	25	1.10″	4.0″
September 28	81	28	1.37″	
September 29	80	28	1.68″	0.3″
September 30	77	22	1.12″	
October 1	81	24	1.59″	0.3″
October 2	84	27	0.73″	2.0″
October 3	80	23	1.65″	
October 4	87	26	1.44″	0.3″
October 5	82	24	2.35″	2.5″
October 6	83	18	1.35″	2.5″
October 7	77	22	1.74″	0.5″
October 8	78	22	1.73″	
October 9	79	23	0.82″	2.0″
October 10	84	22	0.77″	0.5″
October 11	83	26	0.76″	1.9″
October 12	79	26	1.79″	6.0″
October 13	83	19	1.87″	
October 14	78	23	1.30″	1.5″
October 15	78	20	2.02″	1.5″
October 16	78	23	0.90″	0.7″
October 17	84	15	2.00″	2.5″
October 18	83	16	2.00″	8.0″
October 19	78	19	1.24″	6.0″
October 20	78	20	1.44″	5.4″
October 21	80	23	0.90″	9.0″
October 22	78	15	1.32″	9.0″
October 23	82	19	2.10″	3.8″
October 24	74	13	2.64″	6.0″
October 25	75	20	2.80″	3.9″
October 26	78	20	1.06″	10.0″
October 27	75	15	0.66″	2.5″
October 28	67	11	0.80″	4.8″
October 29	70	17	1.50″	3.0″
October 30	75	4	1.15″	2.5″
October 31	74	19	0.93″	5.1″
November 1	69	16	1.34″	4.0″
November 2	69	11	1.03″	2.5″
November 3	70	3	1.16″	3.0″
November 4	65	12	1.36″	6.2″
November 5	67	8	2.14″	7.0″
November 6	68	−1	2.00″	8.0″
November 7	68	3	0.93″	6.0″
November 8	66	6	1.74″	9.0″
November 9	67	10	1.37″	5.0″
November 10	70	12	1.26″	11.8″
November 11	64	0	1.48″	8.0″
November 12	58	6	1.65″	4.5″
November 13	57	4	1.00″	6.0″
November 14	65	11	1.22″	6.0″
November 15	60	−1	1.35″	13.0″
November 16	68	1	1.04″	6.0″
November 17	66	−3	1.34″	7.9″
November 18	68	2	0.87″	7.0″

Munising *(continued)*

	RECORD HIGH	RECORD LOW	RECORD PRECIPITATION	RECORD SNOWFALL
November 19	67	11	0.94″	9.0″
November 20	62	3	1.40″	6.0″
November 21	58	5	1.37″	8.0″
November 22	58	–8	1.11″	10.0″
November 23	56	2	0.84″	6.0″
November 24	59	–5	0.86″	8.3″
November 25	57	–10	0.67″	9.0″
November 26	60	6	1.15″	10.0″
November 27	52	–1	0.56″	6.0″
November 28	51	–7	1.02″	6.0″
November 29	52	0	0.94″	10.0″
November 30	56	–2	0.85″	7.5″
December 1	56	–10	0.68″	7.5″
December 2	56	–8	0.80″	8.0″
December 3	60	–9	0.47″	7.5″
December 4	59	–1	0.65″	8.0″
December 5	51	1	1.18″	6.5″
December 6	54	–7	0.73″	11.0″
December 7	46	–10	0.58″	10.0″
December 8	47	–9	0.85″	12.0″
December 9	52	–15	1.08″	12.0″
December 10	49	–10	0.74″	12.0″
December 11	48	–10	1.40″	9.0″
December 12	48	–8	1.24″	13.9″
December 13	50	–12	0.71″	8.0″
December 14	51	–14	1.00″	9.5″
December 15	42	–18	1.41″	14.0″
December 16	51	–10	1.11″	10.0″
December 17	45	–10	0.70″	7.2″
December 18	47	–16	0.60″	6.0″
December 19	44	–21	1.11″	6.0″
December 20	49	–21	0.58″	6.2″
December 21	45	–12	0.66″	5.0″
December 22	45	–15	0.67″	8.3″
December 23	45	–18	0.88″	8.0″
December 24	43	–18	1.15″	9.0″
December 25	50	–13	0.64″	8.0″
December 26	42	–12	1.32″	9.0″
December 27	45	–12	1.15″	10.0″
December 28	46	–20	0.64″	10.0″
December 29	42	–18	0.69″	7.5″
December 30	50	–21	0.61″	7.0″
December 31	47	–11	1.02″	10.9″

Pontiac (Climate Record Begins 1894)

	RECORD HIGH	RECORD LOW	RECORD PRECIPITATION	RECORD SNOWFALL
January 1	58	–9	1.20″	8.5″
January 2	55	–3	0.73″	5.0″
January 3	56	–7	1.16″	7.3″
January 4	56	–12	0.75″	3.0″
January 5	61	–8	1.17″	4.0″
January 6	56	–6	0.87″	3.5″
January 7	55	–9	0.62″	3.0″
January 8	63	–7	1.00″	4.0″
January 9	59	–4	1.16″	4.0″
January 10	54	–9	0.47″	6.0″
January 11	54	–11	0.83″	5.0″
January 12	53	–18	0.85″	5.0″
January 13	59	–13	0.95″	6.0″
January 14	59	–8	0.99″	8.0″
January 15	54	–14	0.88″	9.0″
January 16	55	–13	0.54″	5.0″
January 17	59	–15	0.58″	4.0″
January 18	58	–11	0.88″	7.2″
January 19	62	–21	0.82″	5.0″
January 20	58	–11	0.65″	3.2″
January 21	64	–15	0.74″	6.0″
January 22	60	–15	0.56″	10.0″
January 23	59	–14	0.92″	6.0″
January 24	61	–15	0.72″	4.0″
January 25	66	–13	0.60″	6.5″
January 26	64	–10	1.05″	3.5″
January 27	62	–12	1.28″	6.0″
January 28	58	–11	0.41″	3.6″
January 29	54	–9	1.40″	5.0″
January 30	51	–18	1.08″	9.0″
January 31	58	–8	1.03″	6.0″
February 1	57	–13	1.02″	12.0″
February 2	52	–12	1.29″	4.2″
February 3	54	–11	0.45″	4.3″
February 4	56	–15	0.60″	6.0″
February 5	59	**–22**	1.44″	5.8″
February 6	54	–16	1.00″	12.0″
February 7	59	–9	0.56″	4.0″
February 8	54	–15	1.25″	4.0″
February 9	57	–20	1.20″	5.0″
February 10	54	–20	1.19″	3.0″
February 11	57	–9	1.40″	8.0″
February 12	58	–21	3.51″	6.0″

Pontiac *(continued)*

	RECORD HIGH	RECORD LOW	RECORD PRECIPITATION	RECORD SNOWFALL
February 13	59	−10	0.96″	4.0″
February 14	57	−7	1.25″	5.0″
February 15	59	−8	0.95″	6.5″
February 16	58	−5	1.36″	6.0″
February 17	55	−10	1.05″	4.0″
February 18	55	−14	0.94″	3.0″
February 19	58	−10	1.31″	7.0″
February 20	60	−20	0.94″	2.0″
February 21	60	−6	0.61″	5.0″
February 22	62	−14	1.08″	4.2″
February 23	65	−7	1.30″	4.0″
February 24	62	−8	0.94″	5.0″
February 25	62	−5	1.50″	7.0″
February 26	64	−6	1.24″	6.0″
February 27	60	−8	0.55″	2.0″
February 28	57	−2	0.89″	4.0″
February 29	55	−4	0.10″	1.4″
March 1	60	−3	0.85″	4.0″
March 2	59	−8	1.61″	4.0″
March 3	65	−8	0.67″	2.0″
March 4	66	2	1.32″	5.8″
March 5	64	−3	1.86″	4.5″
March 6	66	0	1.22″	4.0″
March 7	71	−7	0.90″	6.0″
March 8	73	−5	1.55″	6.0″
March 9	69	−1	1.15″	6.5″
March 10	65	3	0.71″	3.0″
March 11	68	−8	1.08″	5.6″
March 12	74	1	0.80″	3.0″
March 13	73	1	0.59″	3.0″
March 14	75	−4	1.18″	3.0″
March 15	76	2	1.37″	7.0″
March 16	71	1	1.45″	2.0″
March 17	72	0	0.76″	5.5″
March 18	71	5	1.23″	5.2″
March 19	75	−1	1.18″	3.0″
March 20	72	4	0.84″	7.0″
March 21	72	8	1.20″	7.0″
March 22	80	4	1.42″	6.0″
March 23	76	−5	0.91″	3.0″
March 24	81	−4	1.00″	5.3″
March 25	78	0	1.43″	5.0″
March 26	78	6	0.58″	5.0″
March 27	78	7	1.32″	11.0″
March 28	80	−1	0.92″	1.9″
March 29	78	6	0.70″	2.0″
March 30	78	13	1.14″	8.0″
March 31	78	12	1.67″	1.0″
April 1	77	6	1.57″	3.0″
April 2	82	17	1.52″	6.0″
April 3	78	12	1.44″	8.0″
April 4	75	14	1.56″	3.0″
April 5	77	15	2.52″	3.0″
April 6	81	16	1.30″	9.5″
April 7	81	8	1.00″	2.0″
April 8	78	11	0.98″	2.6″
April 9	76	17	0.89″	2.0″
April 10	79	16	1.60″	4.0″
April 11	82	20	1.02″	2.0″
April 12	83	18	0.87″	3.0″
April 13	84	16	0.81″	0.2″
April 14	80	16	1.30″	1.0″
April 15	80	18	1.68″	3.0″
April 16	84	18	0.90″	2.5″
April 17	85	23	1.21″	5.0″
April 18	82	21	1.31″	4.0″
April 19	85	21	2.45″	1.2″
April 20	84	23	1.78″	1.0″
April 21	84	23	1.07″	
April 22	87	22	1.00″	0.1″
April 23	86	21	1.06″	
April 24	85	22	0.94″	1.5″
April 25	89	23	4.00″	1.0″
April 26	88	23	1.03″	
April 27	84	26	1.13″	1.0″
April 28	84	26	1.02″	
April 29	81	28	2.85″	5.0″
April 30	85	24	1.56″	
May 1	85	25	2.30″	0.2″
May 2	84	28	1.55″	0.2″
May 3	87	23	1.17″	
May 4	87	25	0.90″	
May 5	90	30	0.84″	
May 6	87	25	0.89″	
May 7	85	28	1.44″	
May 8	86	28	0.80″	
May 9	87	26	1.08″	9.0″
May 10	88	24	2.55″	
May 11	87	28	1.94″	
May 12	87	29	2.00″	
May 13	87	29	1.29″	0.5″
May 14	89	30	1.32″	
May 15	90	29	1.92″	
May 16	90	31	1.10″	
May 17	92	29	1.67″	
May 18	92	31	1.21″	
May 19	91	30	1.85″	
May 20	90	30	1.00″	
May 21	89	29	1.09″	
May 22	88	34	1.46″	
May 23	86	30	2.19″	

Pontiac *(continued)*

	RECORD HIGH	RECORD LOW	RECORD PRECIPITATION	RECORD SNOWFALL
May 24	87	31	1.21″	
May 25	88	30	1.65″	
May 26	93	34	2.00″	
May 27	95	32	1.97″	
May 28	93	36	1.35″	
May 29	91	33	0.70″	
May 30	93	33	1.31″	
May 31	93	35	1.27″	
June 1	100	35	1.47″	
June 2	94	37	2.06″	
June 3	94	35	1.19″	
June 4	95	36	1.54″	
June 5	95	37	0.95″	
June 6	96	36	1.46″	
June 7	93	40	1.86″	
June 8	94	35	2.12″	
June 9	93	36	1.50″	
June 10	96	34	2.08″	
June 11	95	37	1.94″	
June 12	95	40	3.32″	
June 13	95	42	0.91″	
June 14	97	41	2.54″	
June 15	95	41	1.65″	
June 16	96	38	1.09″	
June 17	95	39	1.39″	
June 18	94	42	3.52″	
June 19	98	40	2.02″	
June 20	96	38	1.61″	
June 21	94	40	1.78″	
June 22	96	40	3.70″	
June 23	97	40	2.30″	
June 24	95	42	0.94″	
June 25	102	43	1.25″	
June 26	99	42	4.70″	
June 27	99	46	1.47″	
June 28	100	45	2.03″	
June 29	98	43	1.65″	
June 30	96	43	1.63″	
July 1	98	45	1.32″	
July 2	100	43	1.26″	
July 3	101	45	1.41″	
July 4	99	43	2.95″	
July 5	103	41	2.91″	
July 6	**104**	45	0.94″	
July 7	101	45	1.00″	
July 8	**104**	45	2.58″	
July 9	102	45	2.00″	
July 10	100	42	1.74″	
July 11	100	44	0.85″	
July 12	100	49	1.49″	
July 13	103	49	1.03″	
July 14	102	47	1.82″	
July 15	97	44	1.01″	
July 16	**104**	47	1.15″	
July 17	100	42	2.06″	
July 18	97	45	2.52″	
July 19	98	46	1.18″	
July 20	102	46	1.20″	
July 21	98	46	1.03″	
July 22	96	49	1.79″	
July 23	101	45	3.38″	
July 24	**104**	48	1.57″	
July 25	99	47	1.69″	
July 26	98	47	3.34″	
July 27	96	48	2.68″	
July 28	99	50	1.95″	
July 29	100	47	1.30″	
July 30	100	48	3.04″	
July 31	98	45	1.78″	
August 1	96	45	1.70″	
August 2	101	43	0.80″	
August 3	101	45	1.96″	
August 4	98	42	1.46″	
August 5	96	44	1.44″	
August 6	102	47	1.33″	
August 7	101	48	1.22″	
August 8	98	48	2.45″	
August 9	94	43	2.45″	
August 10	97	45	2.83″	
August 11	96	45	1.53″	
August 12	95	45	1.78″	
August 13	98	45	1.57″	
August 14	99	45	0.53″	
August 15	97	47	2.02″	
August 16	95	43	2.31″	
August 17	96	41	1.54″	
August 18	95	41	1.27″	
August 19	97	44	1.95″	
August 20	97	41	1.33″	
August 21	99	41	1.89″	
August 22	97	37	1.44″	
August 23	96	40	2.04″	
August 24	95	44	1.17″	
August 25	96	43	2.29″	
August 26	96	46	1.55″	
August 27	99	40	1.53″	
August 28	96	40	2.32″	
August 29	96	40	1.35″	
August 30	97	42	1.33″	
August 31	96	38	0.81″	

Pontiac *(continued)*

	RECORD HIGH	RECORD LOW	RECORD PRECIPITATION	RECORD SNOWFALL
September 1	98	41	2.57″	
September 2	98	41	0.99″	
September 3	97	39	0.72″	
September 4	92	42	2.30″	
September 5	98	43	1.43″	
September 6	95	38	1.06″	
September 7	92	41	2.38″	
September 8	96	39	2.00″	
September 9	96	39	1.28″	
September 10	94	35	1.45″	
September 11	95	33	2.52″	
September 12	94	38	1.87″	
September 13	94	36	1.50″	
September 14	93	33	2.83″	
September 15	97	38	1.73″	
September 16	92	38	0.55″	
September 17	93	35	2.00″	
September 18	92	35	2.76″	
September 19	96	33	0.89″	
September 20	91	33	0.81″	
September 21	93	33	2.61″	
September 22	91	31	1.03″	
September 23	87	29	1.93″	
September 24	87	32	1.17″	
September 25	90	34	1.06″	
September 26	92	30	0.97″	
September 27	88	32	1.70″	
September 28	90	31	2.80″	
September 29	85	30	1.15″	
September 30	86	32	2.09″	
October 1	85	28	**4.75**″	
October 2	85	26	0.92″	
October 3	90	25	1.43″	
October 4	89	28	3.72″	
October 5	88	28	0.88″	
October 6	89	28	1.81″	
October 7	88	26	1.18″	
October 8	90	27	0.51″	
October 9	88	27	0.94″	
October 10	84	26	1.13″	
October 11	83	23	1.24″	
October 12	83	25	0.95″	
October 13	83	23	1.37″	0.6″
October 14	84	22	0.98″	
October 15	84	28	1.56″	
October 16	85	23	1.12″	1.0″
October 17	87	23	1.42″	0.6″
October 18	84	23	1.53″	
October 19	84	22	2.33″	1.5″
October 20	80	23	1.50″	
October 21	81	15	0.74″	

	RECORD HIGH	RECORD LOW	RECORD PRECIPITATION	RECORD SNOWFALL
October 22	81	24	2.00″	1.6″
October 23	82	19	2.00″	
October 24	80	21	2.39″	
October 25	81	22	0.73″	
October 26	81	22	0.65″	0.3″
October 27	81	19	0.96″	
October 28	73	22	0.75″	2.0″
October 29	74	20	0.73″	2.0″
October 30	76	16	1.05″	1.0″
October 31	78	21	0.73″	
November 1	79	20	0.70″	
November 2	75	16	0.96″	1.0″
November 3	76	12	1.45″	3.3″
November 4	72	18	0.53″	0.5″
November 5	73	13	1.78″	0.5″
November 6	73	16	1.15″	2.0″
November 7	70	18	1.20″	7.0″
November 8	71	14	0.80″	4.0″
November 9	70	16	1.37″	8.0″
November 10	68	18	1.40″	2.0″
November 11	69	15	1.60″	3.0″
November 12	70	14	0.90″	0.8″
November 13	70	10	1.12″	3.0″
November 14	69	12	1.18″	4.2″
November 15	68	5	1.08″	4.0″
November 16	70	3	2.07″	4.0″
November 17	70	9	0.71″	6.0″
November 18	67	10	1.17″	2.0″
November 19	69	13	1.27″	4.0″
November 20	67	10	0.94″	2.5″
November 21	67	7	0.70″	2.0″
November 22	64	7	2.10″	3.7″
November 23	67	10	0.65″	1.0″
November 24	67	4	1.00″	4.3″
November 25	61	4	1.05″	3.8″
November 26	69	5	1.01″	3.5″
November 27	65	3	0.99″	4.5″
November 28	65	4	0.93″	4.0″
November 29	70	2	1.05″	4.0″
November 30	65	2	1.20″	3.0″
December 1	62	4	1.48″	2.0″
December 2	63	0	1.60″	**18.0**″
December 3	64	–3	0.82″	3.0″
December 4	62	–1	0.60″	4.0″
December 5	60	–1	1.75″	4.0″
December 6	65	0	0.52″	3.5″
December 7	60	3	1.83″	4.5″
December 8	64	4	1.00″	3.0″
December 9	59	–6	0.67″	8.0″
December 10	63	–2	0.74″	3.3″

Pontiac *(continued)*

	RECORD HIGH	RECORD LOW	RECORD PRECIPITATION	RECORD SNOWFALL
December 11	62	−4	0.85″	10.0″
December 12	59	−2	1.11″	4.0″
December 13	57	−9	1.50″	6.0″
December 14	62	−3	0.70″	5.0″
December 15	59	−7	0.80″	5.0″
December 16	59	−7	0.54″	5.0″
December 17	59	−8	1.58″	6.0″
December 18	52	−7	0.70″	4.0″
December 19	53	−6	1.17″	7.0″
December 20	57	−3	1.10″	9.1″
December 21	58	−8	1.46″	4.0″
December 22	60	−6	0.84″	5.7″
December 23	54	−7	0.92″	5.0″
December 24	54	−9	0.47″	3.2″
December 25	61	−8	2.34″	6.0″
December 26	61	−11	0.80″	5.0″
December 27	56	−11	1.37″	3.8″
December 28	59	−12	1.07″	5.0″
December 29	59	−5	0.90″	5.0″
December 30	52	−4	1.70″	3.0″
December 31	58	−6	1.10″	6.0″

Port Huron (Climate Record Begins 1931)

	RECORD HIGH	RECORD LOW	RECORD PRECIPITATION	RECORD SNOWFALL
January 1	56	−5	1.45″	7.0″
January 2	54	−1	0.65″	9.5″
January 3	56	−4	1.08″	2.5″
January 4	58	−9	1.33″	4.4″
January 5	55	−2	0.57″	6.0″
January 6	54	−6	0.36″	4.5″
January 7	60	−7	0.92″	4.0″
January 8	57	−7	0.60″	2.0″
January 9	57	−3	0.67″	5.0″
January 10	52	−7	0.75″	8.0″
January 11	50	−4	0.60″	5.0″
January 12	53	−2	1.22″	12.3″
January 13	59	−2	0.85″	6.5″
January 14	59	−5	1.38″	12.0″
January 15	52	−11	1.00″	8.0″
January 16	53	−13	0.39″	4.0″
January 17	55	−15	0.80″	5.0″
January 18	58	−13	1.18″	3.5″
January 19	53	**−19**	0.60″	3.4″
January 20	50	−12	1.21″	4.0″
January 21	51	−11	0.62″	4.1″
January 22	59	−7	0.61″	3.5″
January 23	60	−3	0.93″	13.0″
January 24	59	−13	1.40″	3.0″
January 25	64	−5	0.45″	5.0″
January 26	64	−9	1.68″	4.4″
January 27	56	−9	2.30″	7.0″
January 28	54	−7	0.78″	4.5″
January 29	51	−7	0.58″	3.4″
January 30	51	−10	1.00″	9.8″
January 31	57	−7	0.91″	7.0″
February 1	52	−9	1.61″	5.0″
February 2	48	−11	0.83″	7.5″
February 3	52	−8	0.50″	7.0″
February 4	56	−6	0.90″	6.5″
February 5	49	−7	1.45″	3.0″
February 6	54	−6	1.58″	8.0″
February 7	49	−6	0.62″	5.6″
February 8	54	−10	0.48″	5.0″
February 9	55	−15	1.37″	2.0″
February 10	56	−12	1.53″	6.0″
February 11	64	−4	1.80″	8.0″
February 12	60	−5	1.12″	7.5″
February 13	58	−3	0.68″	5.0″
February 14	54	−2	1.33″	6.0″
February 15	58	−7	0.82″	4.0″
February 16	51	−5	1.56″	4.0″
February 17	52	−8	1.20″	3.0″
February 18	57	−10	0.55″	5.5″
February 19	59	−6	1.60″	4.4″
February 20	59	−7	0.61″	5.3″
February 21	58	−3	0.94″	4.0″
February 22	53	−3	1.02″	4.0″
February 23	60	1	1.20″	10.0″
February 24	52	−6	1.06″	4.0″
February 25	61	−1	1.02″	13.0″
February 26	69	−6	0.80″	8.0″
February 27	61	−6	0.75″	1.5″
February 28	53	4	1.12″	2.7″
February 29	55	1	0.13″	1.0″
March 1	62	−2	1.12″	3.0″
March 2	58	2	0.60″	5.0″
March 3	64	−7	0.55″	2.0″
March 4	59	−6	1.80″	8.0″
March 5	64	0	1.11″	6.0″
March 6	69	3	0.95″	5.0″

Port Huron *(continued)*

	RECORD HIGH	RECORD LOW	RECORD PRECIPITATION	RECORD SNOWFALL
March 7	76	2	1.01″	6.3″
March 8	79	–6	0.85″	9.1″
March 9	67	1	1.33″	5.4″
March 10	63	4	0.69″	2.5″
March 11	67	3	0.95″	3.5″
March 12	74	–1	1.02″	4.0″
March 13	74	10	0.64″	2.0″
March 14	73	5	1.12″	6.5″
March 15	74	6	0.80″	1.0″
March 16	72	8	1.25″	3.0″
March 17	70	8	1.44″	7.0″
March 18	75	3	1.32″	2.5″
March 19	70	6	1.14″	2.7″
March 20	67	10	1.42″	10.0″
March 21	69	8	1.20″	5.0″
March 22	77	9	0.91″	6.3″
March 23	71	0	1.03″	12.0″
March 24	75	8	0.79″	3.0″
March 25	77	4	1.02″	6.2″
March 26	78	10	0.46″	4.0″
March 27	78	6	1.10″	**14.3**″
March 28	80	8	0.78″	0.0″
March 29	76	14	1.01″	4.5″
March 30	80	11	1.65″	1.5″
March 31	79	15	1.18″	1.3″
April 1	78	19	0.90″	5.4″
April 2	82	14	1.25″	6.0″
April 3	81	8	1.18″	10.0″
April 4	69	15	1.60″	1.0″
April 5	73	13	1.01″	5.7″
April 6	80	17	1.23″	2.3″
April 7	85	15	0.57″	0.7″
April 8	74	14	0.81″	2.0″
April 9	78	19	0.73″	1.3″
April 10	82	15	0.74″	3.5″
April 11	83	20	0.93″	2.0″
April 12	83	18	1.35″	3.4″
April 13	87	16	1.40″	1.2″
April 14	78	19	1.26″	
April 15	82	19	0.80″	
April 16	85	24	1.42″	3.2″
April 17	84	24	1.56″	3.0″
April 18	84	22	1.90″	
April 19	79	21	1.38″	
April 20	84	21	1.95″	
April 21	85	25	0.86″	
April 22	85	21	1.42″	
April 23	85	21	0.90″	1.0″
April 24	87	27	0.74″	0.6″
April 25	87	25	1.62″	1.0″
April 26	87	26	1.12″	0.5″
April 27	86	28	0.97″	
April 28	85	26	1.90″	
April 29	82	27	1.30″	
April 30	85	28	0.97″	
May 1	85	28	0.85″	
May 2	83	26	1.20″	
May 3	88	28	1.54″	
May 4	89	28	0.80″	
May 5	91	30	0.76″	
May 6	91	30	1.53″	
May 7	89	26	2.46″	
May 8	87	27	0.94″	
May 9	88	29	1.19″	
May 10	85	21	1.79″	
May 11	87	29	3.39″	
May 12	86	29	1.18″	
May 13	89	32	0.95″	
May 14	90	30	1.20″	
May 15	93	36	1.60″	
May 16	93	31	1.40″	
May 17	93	33	1.06″	
May 18	92	31	1.68″	
May 19	95	30	1.40″	
May 20	91	32	1.25″	
May 21	92	32	1.34″	
May 22	90	32	1.00″	
May 23	90	32	2.98″	
May 24	89	32	1.90″	
May 25	89	35	1.00″	
May 26	87	36	2.18″	
May 27	90	34	0.83″	
May 28	93	38	0.77″	
May 29	96	35	1.08″	
May 30	95	37	0.71″	
May 31	95	38	0.82″	
June 1	94	39	1.22″	
June 2	93	40	2.98″	
June 3	88	38	0.85″	
June 4	92	35	1.04″	
June 5	93	32	2.17″	
June 6	95	35	1.43″	
June 7	93	37	1.59″	
June 8	96	33	1.26″	
June 9	91	33	2.07″	
June 10	95	34	0.87″	
June 11	96	35	2.41″	
June 12	95	45	1.07″	
June 13	95	44	1.51″	
June 14	97	37	1.67″	
June 15	95	41	1.75″	

Port Huron *(continued)*

	RECORD HIGH	RECORD LOW	RECORD PRECIPITATION	RECORD SNOWFALL
June 16	98	41	2.11″	
June 17	93	42	2.61″	
June 18	98	40	2.26″	
June 19	98	45	1.80″	
June 20	98	41	1.67″	
June 21	94	37	1.39″	
June 22	93	37	2.51″	
June 23	92	43	1.35″	
June 24	97	43	1.77″	
June 25	102	45	1.52″	
June 26	98	46	1.99″	
June 27	97	44	0.75″	
June 28	100	51	1.60″	
June 29	100	47	1.00″	
June 30	96	43	0.98″	
July 1	97	35	1.51″	
July 2	97	46	0.88″	
July 3	98	44	1.06″	
July 4	98	44	1.30″	
July 5	99	42	0.74″	
July 6	100	47	0.98″	
July 7	100	48	1.99″	
July 8	99	45	1.40″	
July 9	**103**	49	1.65″	
July 10	102	50	1.75″	
July 11	100	50	1.64″	
July 12	97	39	1.31″	
July 13	97	44	0.89″	
July 14	102	42	1.49″	
July 15	100	47	1.30″	
July 16	96	40	2.90″	
July 17	97	44	3.57″	
July 18	97	51	1.58″	
July 19	95	51	1.22″	
July 20	98	45	1.25″	
July 21	95	49	1.78″	
July 22	95	51	3.72″	
July 23	97	47	0.90″	
July 24	100	48	1.85″	
July 25	100	45	1.06″	
July 26	95	43	0.98″	
July 27	96	50	1.50″	
July 28	97	50	1.19″	
July 29	96	50	0.94″	
July 30	96	49	1.32″	
July 31	102	49	1.40″	
August 1	94	47	0.85″	
August 2	101	49	1.06″	
August 3	99	49	1.22″	
August 4	97	49	3.04″	
August 5	96	43	0.99″	
August 6	94	47	1.49″	
August 7	98	43	2.74″	
August 8	97	43	2.17″	
August 9	102	49	2.18″	
August 10	98	47	2.28″	
August 11	100	46	1.80″	
August 12	91	44	0.66″	
August 13	96	45	1.00″	
August 14	100	41	1.27″	
August 15	96	48	1.97″	
August 16	93	45	1.22″	
August 17	96	47	1.24″	
August 18	93	47	1.32″	
August 19	97	44	1.00″	
August 20	98	40	1.19″	
August 21	102	40	1.06″	
August 22	100	43	1.72″	
August 23	95	44	1.11″	
August 24	96	38	2.24″	
August 25	98	37	1.85″	
August 26	97	39	1.31″	
August 27	98	43	2.21″	
August 28	98	41	1.94″	
August 29	97	41	1.74″	
August 30	101	43	2.93″	
August 31	96	46	1.15″	
September 1	100	43	1.08″	
September 2	101	43	0.88″	
September 3	100	42	3.02″	
September 4	96	45	0.99″	
September 5	92	43	0.95″	
September 6	94	39	1.35″	
September 7	95	40	**3.97**″	
September 8	97	37	1.89″	
September 9	96	35	1.34″	
September 10	95	41	1.32″	
September 11	98	38	1.82″	
September 12	94	31	1.12″	
September 13	95	37	1.75″	
September 14	89	35	1.99″	
September 15	93	36	1.40″	
September 16	91	39	1.28″	
September 17	92	37	0.95″	
September 18	94	30	2.01″	
September 19	95	30	1.94″	
September 20	91	35	1.05″	
September 21	91	34	1.25″	
September 22	92	35	2.07″	
September 23	87	33	1.71″	
September 24	85	29	1.03″	

Port Huron *(continued)*

	RECORD HIGH	RECORD LOW	RECORD PRECIPITATION	RECORD SNOWFALL
September 25	90	32	1.45″	
September 26	91	34	0.85″	
September 27	90	32	0.92″	
September 28	86	32	0.88″	
September 29	85	25	1.58″	
September 30	82	32	0.89″	
October 1	86	33	2.00″	
October 2	86	33	0.95″	
October 3	85	32	1.71″	
October 4	86	29	1.07″	
October 5	90	31	2.87″	
October 6	90	26	1.90″	
October 7	89	27	1.56″	
October 8	90	25	0.96″	
October 9	86	28	1.56″	
October 10	83	30	1.07″	
October 11	81	26	1.40″	
October 12	82	27	1.33″	
October 13	81	28	1.47″	
October 14	82	25	1.15″	
October 15	82	28	1.07″	
October 16	83	25	1.46″	
October 17	83	24	1.00″	
October 18	84	24	1.48″	
October 19	86	24	0.89″	2.0″
October 20	77	24	0.60″	6.0″
October 21	81	24	1.57″	
October 22	79	25	1.13″	
October 23	81	26	1.28″	0.5″
October 24	80	22	1.68″	
October 25	79	23	0.48″	
October 26	81	25	0.81″	2.0″
October 27	80	20	2.03″	4.5″
October 28	74	22	0.65″	0.5″
October 29	74	22	0.92″	
October 30	76	21	0.90″	0.6″
October 31	76	20	0.95″	
November 1	81	21	0.88″	
November 2	74	25	1.37″	
November 3	75	13	1.29″	
November 4	72	15	1.06″	4.0″
November 5	71	13	2.06″	3.5″
November 6	72	12	1.42″	2.0″
November 7	67	18	1.58″	7.0″
November 8	70	14	1.42″	
November 9	75	18	1.08″	5.0″
November 10	68	19	2.26″	1.0″
November 11	67	17	1.28″	3.1″
November 12	66	17	2.00″	3.2″
November 13	67	12	0.64″	2.0″
November 14	65	12	1.54″	2.0″
November 15	68	6	1.87″	3.5″
November 16	68	5	1.20″	5.4″
November 17	67	11	0.88″	1.0″
November 18	71	11	0.47″	1.5″
November 19	67	16	1.56″	3.0″
November 20	69	15	1.43″	4.0″
November 21	67	12	0.86″	1.5″
November 22	61	13	0.80″	1.5″
November 23	69	13	0.64″	2.0″
November 24	64	2	0.81″	3.0″
November 25	61	10	3.06″	8.5″
November 26	59	14	0.78″	4.0″
November 27	66	8	0.89″	12.0″
November 28	66	9	1.02″	2.5″
November 29	64	7	0.90″	6.0″
November 30	64	4	1.75″	4.4″
December 1	63	8	0.86″	5.0″
December 2	64	1	1.10″	11.0″
December 3	66	0	0.68″	6.2″
December 4	63	7	0.75″	1.3″
December 5	60	7	1.49″	7.0″
December 6	66	6	0.90″	13.0″
December 7	61	6	1.15″	4.0″
December 8	61	4	0.96″	8.0″
December 9	60	2	0.66″	2.4″
December 10	63	2	0.94″	7.0″
December 11	59	2	1.52″	12.0″
December 12	59	1	1.16″	12.5″
December 13	52	3	2.00″	5.5″
December 14	61	3	0.96″	4.0″
December 15	59	0	0.77″	7.0″
December 16	60	−5	0.79″	8.0″
December 17	55	−5	0.58″	5.8″
December 18	51	−1	1.70″	7.0″
December 19	53	−4	0.80″	6.8″
December 20	55	−6	1.00″	3.8″
December 21	55	−6	1.05″	13.0″
December 22	57	−6	1.11″	3.0″
December 23	52	−7	0.57″	10.0″
December 24	57	−7	1.21″	2.0″
December 25	62	−2	2.00″	5.0″
December 26	52	−5	0.91″	5.0″
December 27	58	−5	1.15″	3.5″
December 28	61	4	1.12″	3.0″
December 29	63	−2	0.89″	4.0″
December 30	50	−2	0.99″	5.0″
December 31	56	−3	1.20″	2.3″

Sault Ste. Marie (Climate Record Begins 1888)

	RECORD HIGH	RECORD LOW	RECORD PRECIPITATION	RECORD SNOWFALL
January 1	40	–23	0.49″	5.0″
January 2	43	–22	1.02″	10.0″
January 3	44	–28	1.61″	5.4″
January 4	44	–32	1.07″	7.2″
January 5	44	–25	1.21″	6.9″
January 6	43	–26	0.63″	7.0″
January 7	54	–20	0.58″	9.8″
January 8	41	–28	0.56″	4.0″
January 9	40	–30	0.54″	7.0″
January 10	41	–36	0.60″	8.4″
January 11	44	–29	0.44″	8.5″
January 12	43	–30	1.05″	12.0″
January 13	46	–26	0.54″	7.4″
January 14	43	–27	1.16″	6.6″
January 15	41	–29	0.62″	5.8″
January 16	43	–26	0.74″	11.1″
January 17	44	–24	0.59″	7.6″
January 18	45	–26	0.68″	7.7″
January 19	42	–27	0.66″	9.8″
January 20	46	–23	0.85″	6.2″
January 21	43	–33	0.65″	6.5″
January 22	43	–22	0.84″	7.0″
January 23	42	–25	0.62″	6.6″
January 24	44	–19	0.95″	6.9″
January 25	42	–25	1.11″	12.2″
January 26	43	–32	0.45″	7.0″
January 27	44	–28	0.66″	9.2″
January 28	40	–28	0.99″	9.9″
January 29	48	–26	0.95″	11.7″
January 30	40	–25	0.57″	6.4″
January 31	40	–28	0.70″	11.0″
February 1	39	–31	0.66″	7.8″
February 2	39	–28	0.86″	12.4″
February 3	46	–27	0.32″	3.4″
February 4	46	–25	0.34″	10.4″
February 5	40	–31	0.67″	5.8″
February 6	44	–28	0.86″	5.1″
February 7	41	–26	0.39″	5.2″
February 8	44	**–37**	0.79″	6.6″
February 9	41	–35	0.48″	4.6″
February 10	46	**–37**	0.79″	7.3″
February 11	49	–34	0.84″	10.9″
February 12	47	–28	0.54″	7.8″
February 13	42	–27	0.44″	5.1″
February 14	42	–21	0.49″	5.2″
February 15	44	–25	0.88″	9.4″
February 16	50	–25	1.10″	5.7″
February 17	44	–35	0.65″	6.6″
February 18	46	–27	0.59″	8.3″
February 19	46	–23	0.90″	4.6″
February 20	47	–27	0.72″	6.2″
February 21	42	–17	0.54″	5.2″
February 22	44	–23	0.44″	5.8″
February 23	47	–24	0.64″	9.0″
February 24	49	–26	1.00″	4.6″
February 25	46	–30	0.55″	5.0″
February 26	46	–22	0.64″	6.0″
February 27	46	–24	0.66″	5.6″
February 28	46	–17	0.78″	3.5″
February 29	49	–17	0.24″	6.2″
March 1	45	–18	0.47″	4.5″
March 2	46	–17	0.64″	7.8″
March 3	50	–24	1.08″	4.9″
March 4	50	–18	1.15″	9.8″
March 5	49	–19	0.76″	11.8″
March 6	51	–27	0.64″	7.6″
March 7	60	–20	0.78″	8.6″
March 8	59	–21	0.62″	4.2″
March 9	49	–16	1.16″	8.8″
March 10	50	–16	0.73″	6.0″
March 11	47	–20	0.75″	6.5″
March 12	55	–16	0.68″	9.6″
March 13	58	–16	0.91″	4.0″
March 14	62	–15	1.10″	10.1″
March 15	52	–15	1.38″	8.8″
March 16	56	–14	0.54″	4.5″
March 17	56	–16	0.38″	6.0″
March 18	57	–14	0.62″	4.3″
March 19	57	–13	1.14″	6.7″
March 20	61	–15	1.10″	5.8″
March 21	57	–9	1.01″	7.2″
March 22	57	–12	0.97″	8.1″
March 23	60	–14	0.94″	3.8″
March 24	64	–7	1.01″	8.5″
March 25	60	–11	1.13″	4.8″
March 26	65	–9	1.50″	11.4″
March 27	69	–6	0.88″	3.4″
March 28	75	–11	0.74″	7.1″
March 29	60	–4	0.74″	6.3″
March 30	63	0	0.94″	3.8″
March 31	63	–6	1.08″	6.0″
April 1	67	–13	0.68″	5.3″
April 2	58	7	0.87″	6.2″
April 3	64	4	1.26″	4.5″
April 4	67	2	0.89″	8.3″
April 5	75	–2	1.22″	3.9″
April 6	71	6	1.53″	9.0″
April 7	75	2	0.74″	8.3″
April 8	65	1	1.63″	4.7″

Sault Ste. Marie *(continued)*

	RECORD HIGH	RECORD LOW	RECORD PRECIPITATION	RECORD SNOWFALL
April 9	72	–5	1.19″	6.0″
April 10	76	–2	0.96″	3.1″
April 11	79	4	0.66″	2.5″
April 12	73	–1	1.45″	4.7″
April 13	66	4	1.24″	6.2″
April 14	72	11	0.90″	3.3″
April 15	71	11	1.02″	5.1″
April 16	83	14	0.85″	4.2″
April 17	76	17	0.84″	2.6″
April 18	83	15	0.99″	1.5″
April 19	81	5	1.00″	6.2″
April 20	81	10	1.06″	4.0″
April 21	76	16	1.46″	3.0″
April 22	82	20	0.60″	3.2″
April 23	76	20	0.96″	3.0″
April 24	83	13	0.74″	2.1″
April 25	85	18	1.09″	4.7″
April 26	84	18	2.35″	3.0″
April 27	84	19	1.18″	1.5″
April 28	77	20	2.12″	2.2″
April 29	77	22	0.62″	1.5″
April 30	79	20	1.22″	3.2″
May 1	82	21	1.37″	4.5″
May 2	82	21	0.95″	1.1″
May 3	82	22	0.85″	2.7″
May 4	84	20	1.37″	0.7″
May 5	83	23	1.50″	3.8″
May 6	82	22	1.43″	
May 7	84	18	0.88″	0.7″
May 8	88	22	2.11″	1.4″
May 9	87	23	0.98″	0.4″
May 10	87	23	0.99″	0.8″
May 11	84	25	1.18″	1.2″
May 12	82	26	1.33″	1.5″
May 13	84	25	2.34″	
May 14	82	24	1.40″	4.5″
May 15	89	28	1.10″	0.3″
May 16	87	24	0.99″	1.0″
May 17	88	25	0.74″	0.1″
May 18	84	26	0.73″	1.0″
May 19	90	23	0.93″	1.5″
May 20	88	26	1.26″	
May 21	86	25	1.07″	0.1″
May 22	86	27	1.02″	0.6″
May 23	89	28	1.12″	1.8″
May 24	87	27	1.74″	0.1″
May 25	85	28	1.55″	
May 26	82	24	1.73″	0.2″
May 27	90	27	1.11″	2.2″
May 28	89	30	2.40″	0.4″
May 29	89	30	1.22″	4.6″

	RECORD HIGH	RECORD LOW	RECORD PRECIPITATION	RECORD SNOWFALL
May 30	91	31	1.48″	
May 31	91	30	5.08″	
June 1	92	32	1.48″	
June 2	88	28	0.97″	
June 3	88	26	0.73″	
June 4	89	30	1.61″	
June 5	89	30	1.90″	
June 6	89	32	1.93″	
June 7	92	32	1.25″	
June 8	92	28	1.64″	
June 9	88	30	1.75″	
June 10	89	28	1.19″	
June 11	87	31	1.76″	
June 12	90	34	1.91″	
June 13	90	36	1.36″	
June 14	90	32	1.17″	
June 15	90	35	1.51″	
June 16	89	30	1.95″	
June 17	92	33	1.46″	
June 18	92	32	1.04″	
June 19	90	36	0.72″	
June 20	92	32	1.69″	
June 21	89	31	0.98″	
June 22	90	31	2.37″	
June 23	92	37	1.92″	
June 24	89	33	0.97″	
June 25	88	35	1.75″	
June 26	93	35	2.39″	
June 27	93	36	1.94″	
June 28	89	38	1.31″	
June 29	90	37	1.78″	
June 30	93	36	2.00″	
July 1	97	37	1.69″	
July 2	97	38	1.52″	
July 3	**98**	38	0.95″	
July 4	94	37	1.66″	
July 5	94	39	1.25″	
July 6	96	36	1.54″	
July 7	97	40	1.62″	
July 8	97	40	1.82″	
July 9	96	41	1.02″	
July 10	93	38	1.01″	
July 11	96	36	0.84″	
July 12	96	40	1.88″	
July 13	92	39	2.20″	
July 14	92	36	1.24″	
July 15	92	37	2.23″	
July 16	93	40	0.95″	
July 17	91	38	1.21″	
July 18	93	42	1.40″	

Sault Ste. Marie *(continued)*

	RECORD HIGH	RECORD LOW	RECORD PRECIPITATION	RECORD SNOWFALL
July 19	94	40	1.51″	
July 20	91	40	1.23″	
July 21	92	36	1.80″	
July 22	90	42	1.58″	
July 23	92	39	1.13″	
July 24	92	42	1.10″	
July 25	91	41	1.54″	
July 26	92	40	1.34″	
July 27	92	39	1.23″	
July 28	90	39	1.46″	
July 29	96	40	1.59″	
July 30	**98**	40	0.98″	
July 31	97	40	1.75″	
August 1	96	36	1.75″	
August 2	93	38	1.56″	
August 3	91	38	**5.92″**	
August 4	90	39	1.33″	
August 5	**98**	39	1.47″	
August 6	**98**	38	1.64″	
August 7	92	40	2.69″	
August 8	90	41	1.51″	
August 9	93	41	1.69″	
August 10	90	38	1.68″	
August 11	91	40	2.06″	
August 12	93	38	1.64″	
August 13	91	39	2.16″	
August 14	90	37	0.96″	
August 15	90	35	2.44″	
August 16	88	37	1.26″	
August 17	89	33	0.85″	
August 18	93	35	1.58″	
August 19	92	38	1.54″	
August 20	94	34	1.46″	
August 21	91	37	1.05″	
August 22	85	32	2.19″	
August 23	96	38	1.18″	
August 24	92	34	1.34″	
August 25	90	35	1.63″	
August 26	92	38	0.90″	
August 27	93	35	1.44″	
August 28	93	37	1.94″	
August 29	92	29	1.40″	
August 30	88	35	3.15″	
August 31	89	37	1.06″	
September 1	93	36	1.71″	
September 2	87	31	2.55″	
September 3	92	33	1.60″	
September 4	89	36	5.27″	
September 5	88	34	1.58″	
September 6	90	34	1.33″	

	RECORD HIGH	RECORD LOW	RECORD PRECIPITATION	RECORD SNOWFALL
September 7	89	35	1.48″	
September 8	95	34	1.32″	
September 9	91	29	1.54″	
September 10	92	30	1.17″	
September 11	89	32	2.13″	
September 12	91	32	2.20″	
September 13	86	32	1.45″	0.3″
September 14	89	32	1.66″	
September 15	89	32	1.81″	
September 16	81	29	1.51″	
September 17	86	27	1.03″	
September 18	86	30	1.21″	
September 19	84	28	1.18″	
September 20	85	30	0.91″	2.7″
September 21	84	27	2.14″	
September 22	85	28	1.11″	
September 23	84	25	0.94″	0.1″
September 24	88	28	1.37″	0.1″
September 25	83	25	1.48″	0.3″
September 26	84	27	1.52″	1.5″
September 27	84	26	1.97″	0.2″
September 28	80	27	1.05″	0.6″
September 29	80	26	1.17″	
September 30	84	25	0.89″	0.2″
October 1	80	28	1.55″	0.8″
October 2	81	25	0.56″	0.5″
October 3	77	23	1.21″	0.5″
October 4	82	25	1.26″	0.3″
October 5	81	24	1.50″	1.0″
October 6	80	24	1.65″	1.2″
October 7	76	26	1.08″	0.2″
October 8	77	25	1.08″	0.7″
October 9	79	22	1.40″	1.0″
October 10	76	22	1.26″	0.9″
October 11	78	22	1.54″	0.3″
October 12	81	22	1.30″	0.6″
October 13	78	23	1.10″	0.6″
October 14	78	20	1.42″	1.4″
October 15	80	21	1.11″	0.3″
October 16	80	23	1.34″	4.5″
October 17	77	22	1.09″	10.1″
October 18	77	17	2.24″	6.8″
October 19	78	21	1.13″	6.0″
October 20	76	20	1.28″	7.0″
October 21	70	18	0.97″	3.0″
October 22	76	20	1.12″	3.5″
October 23	77	20	1.39″	1.8″
October 24	79	16	1.86″	3.3″
October 25	67	20	1.24″	2.2″
October 26	75	18	0.78″	3.7″
October 27	72	16	0.98″	7.0″

Sault Ste. Marie *(continued)*

	RECORD HIGH	RECORD LOW	RECORD PRECIPITATION	RECORD SNOWFALL
October 28	64	16	0.86″	2.5″
October 29	69	17	1.23″	2.0″
October 30	74	15	1.31″	2.5″
October 31	68	20	1.24″	4.0″
November 1	66	19	1.65″	4.6″
November 2	67	12	1.44″	6.3″
November 3	74	8	0.90″	2.7″
November 4	67	12	1.11″	6.6″
November 5	68	7	2.33″	5.1″
November 6	64	10	1.02″	4.6″
November 7	62	7	0.84″	3.3″
November 8	60	9	1.20″	6.0″
November 9	67	10	1.25″	4.5″
November 10	62	9	1.60″	3.8″
November 11	62	10	1.34″	5.2″
November 12	61	10	1.13″	4.2″
November 13	58	4	0.84″	5.9″
November 14	60	–1	1.34″	4.5″
November 15	61	–2	1.06″	6.7″
November 16	60	3	1.29″	14.3″
November 17	66	4	1.02″	12.5″
November 18	62	6	0.89″	7.1″
November 19	64	–7	0.60″	8.0″
November 20	58	0	1.21″	3.3″
November 21	63	–1	1.52″	10.0″
November 22	56	–3	1.79″	7.0″
November 23	59	–5	1.48″	6.8″
November 24	62	–4	1.19″	5.3″
November 25	53	0	0.79″	5.7″
November 26	55	–5	1.11″	7.0″
November 27	53	–5	1.13″	7.5″
November 28	55	–9	1.35″	7.1″
November 29	54	–12	1.12″	5.0″
November 30	59	–10	0.61″	4.5″
December 1	54	–7	0.96″	11.0″
December 2	55	–24	0.90″	8.0″
December 3	60	–14	0.90″	8.9″
December 4	59	–9	0.96″	7.7″
December 5	62	–4	0.81″	6.8″
December 6	58	–10	0.83″	7.6″
December 7	45	–16	0.90″	8.8″
December 8	46	–8	0.85″	8.8″
December 9	51	–20	1.46″	19.4″
December 10	47	–14	1.36″	**26.6″**
December 11	46	–14	0.78″	11.0″
December 12	47	–17	0.94″	8.4″
December 13	45	–19	0.60″	6.9″
December 14	53	–18	0.94″	8.7″
December 15	44	–14	0.90″	7.0″
December 16	49	–18	0.80″	7.2″
December 17	42	–22	0.56″	8.0″
December 18	44	–20	0.84″	5.8″
December 19	45	–20	0.49″	6.1″
December 20	43	–14	0.49″	8.9″
December 21	47	–22	0.83″	5.7″
December 22	45	–10	0.49″	4.5″
December 23	46	–14	0.69″	8.5″
December 24	46	–18	0.89″	7.6″
December 25	47	–25	0.74″	8.0″
December 26	43	–31	0.68″	9.9″
December 27	44	–20	0.64″	6.5″
December 28	45	–19	0.87″	7.6″
December 29	41	–24	0.56″	6.5″
December 30	45	–20	0.61″	5.0″
December 31	48	–17	0.65″	10.0″

Traverse City (Climate Record Begins 1896)

	RECORD HIGH	RECORD LOW	RECORD PRECIPITATION	RECORD SNOWFALL
January 1	55	–7	1.00″	10.0″
January 2	55	–8	0.86″	7.0″
January 3	54	–13	0.78″	7.5″
January 4	49	–8	1.76″	12.2″
January 5	49	–11	0.68″	7.0″
January 6	51	–10	0.99″	9.0″
January 7	50	–17	1.12″	12.0″
January 8	49	–18	0.98″	7.0″
January 9	51	–8	0.50″	5.6″
January 10	51	–11	0.84″	6.3″
January 11	53	–15	0.88″	5.7″
January 12	50	–21	1.02″	6.0″
January 13	49	–12	0.90″	9.0″
January 14	48	–16	0.60″	5.4″
January 15	52	–19	0.50″	8.1″
January 16	53	–15	1.40″	4.7″
January 17	49	–11	1.03″	6.6″
January 18	56	–13	0.71″	5.0″
January 19	50	–21	0.71″	13.7″
January 20	55	–12	0.98″	9.3″

Traverse City *(continued)*

	RECORD HIGH	RECORD LOW	RECORD PRECIPITATION	RECORD SNOWFALL
January 21	50	–20	0.46″	6.0″
January 22	49	–10	1.20″	12.0″
January 23	48	–12	0.90″	6.9″
January 24	54	–10	1.82″	5.0″
January 25	55	–15	1.59″	**16.0″**
January 26	59	–14	0.67″	6.8″
January 27	55	–19	1.23″	14.0″
January 28	55	–14	0.49″	6.0″
January 29	50	–11	0.90″	6.0″
January 30	47	–20	2.10″	13.9″
January 31	49	–20	0.40″	4.0″
February 1	45	–26	1.00″	10.0″
February 2	44	–23	0.60″	8.0″
February 3	50	–25	0.80″	4.2″
February 4	50	–21	0.41″	5.0″
February 5	55	–27	1.00″	6.0″
February 6	50	–22	0.75″	10.3″
February 7	50	–22	1.00″	10.0″
February 8	52	–18	0.86″	8.5″
February 9	50	–27	0.74″	4.0″
February 10	57	–33	0.88″	11.0″
February 11	60	–32	0.70″	8.0″
February 12	54	–29	0.73″	7.3″
February 13	50	–21	0.80″	8.0″
February 14	49	–12	0.84″	15.2″
February 15	52	–18	0.55″	7.0″
February 16	54	–23	0.48″	6.4″
February 17	48	**–37**	0.61″	6.0″
February 18	57	–26	0.55″	5.0″
February 19	54	–20	0.69″	6.0″
February 20	57	–20	1.19″	5.0″
February 21	57	–14	0.80″	8.0″
February 22	57	–13	1.25″	12.5″
February 23	60	–15	0.66″	6.0″
February 24	50	–14	0.74″	5.5″
February 25	52	–22	0.60″	6.0″
February 26	65	–24	0.72″	10.0″
February 27	52	–16	0.72″	4.0″
February 28	52	–22	0.52″	3.6″
February 29	64	–17	0.20″	3.0″
March 1	54	–27	0.66″	3.7″
March 2	53	–28	0.81″	5.0″
March 3	64	–30	1.02″	9.0″
March 4	52	–13	1.20″	12.0″
March 5	64	–13	0.97″	7.5″
March 6	68	–16	1.03″	11.0″
March 7	73	–12	0.65″	7.5″
March 8	77	–11	0.81″	6.0″
March 9	58	–17	1.50″	15.0″
March 10	69	–12	0.69″	8.0″
March 11	71	–10	0.80″	8.0″
March 12	70	–9	0.79″	3.3″
March 13	68	–7	0.60″	4.4″
March 14	73	–6	0.66″	4.0″
March 15	78	–6	1.57″	13.1″
March 16	64	–12	1.03″	2.2″
March 17	74	–8	1.05″	9.5″
March 18	72	–3	1.30″	8.0″
March 19	77	–6	1.20″	6.5″
March 20	71	–10	0.80″	5.0″
March 21	74	–6	0.56″	4.5″
March 22	74	–2	1.35″	5.1″
March 23	70	–4	1.24″	6.0″
March 24	71	–9	0.60″	5.3″
March 25	78	–8	2.00″	5.0″
March 26	75	–7	1.51″	9.0″
March 27	81	–7	0.96″	3.0″
March 28	78	–1	0.83″	4.0″
March 29	82	2	0.50″	3.0″
March 30	81	6	1.37″	6.0″
March 31	76	–2	1.37″	6.0″
April 1	73	3	1.40″	14.0″
April 2	81	1	1.24″	8.0″
April 3	73	3	1.10″	5.0″
April 4	78	3	1.15″	3.6″
April 5	79	6	1.38″	3.2″
April 6	80	4	1.32″	4.0″
April 7	82	11	1.39″	2.8″
April 8	73	3	1.14″	3.0″
April 9	75	9	1.16″	2.1″
April 10	80	13	1.21″	3.0″
April 11	81	16	1.29″	7.2″
April 12	83	16	3.18″	3.0″
April 13	79	11	1.66″	2.0″
April 14	81	17	1.22″	4.0″
April 15	85	18	1.79″	6.0″
April 16	88	6	1.09″	4.0″
April 17	82	19	1.22″	3.5″
April 18	85	20	1.10″	7.0″
April 19	84	15	1.31″	0.5″
April 20	87	16	1.40″	2.0″
April 21	85	16	1.10″	2.0″
April 22	88	19	0.45″	0.4″
April 23	87	21	0.87″	3.2″
April 24	85	22	0.80″	0.9″
April 25	88	20	1.33″	6.0″
April 26	87	22	1.29″	3.8″
April 27	88	22	0.85″	0.2″
April 28	88	21	1.50″	
April 29	90	21	2.30″	3.0″
April 30	85	23	1.46″	2.6″

Traverse City *(continued)*

	RECORD HIGH	RECORD LOW	RECORD PRECIPITATION	RECORD SNOWFALL
May 1	85	21	1.02″	2.0″
May 2	88	17	1.41″	3.0″
May 3	88	22	0.65″	
May 4	85	20	1.08″	2.0″
May 5	90	24	0.97″	0.5″
May 6	89	20	1.45″	0.2″
May 7	85	24	1.13″	
May 8	84	20	0.85″	0.6″
May 9	90	24	1.60″	
May 10	89	22	1.64″	
May 11	88	24	1.84″	
May 12	86	24	0.92″	0.3″
May 13	87	25	1.44″	
May 14	88	26	1.35″	6.0″
May 15	93	27	1.41″	
May 16	93	27	0.76″	
May 17	91	25	0.78″	
May 18	91	26	0.80″	
May 19	92	24	1.40″	
May 20	95	28	1.30″	
May 21	93	26	1.00″	
May 22	89	25	1.43″	
May 23	90	24	2.05″	
May 24	87	25	1.13″	
May 25	87	28	1.00″	
May 26	89	27	1.00″	
May 27	93	28	1.84″	2.0″
May 28	95	29	0.68″	
May 29	93	31	4.18″	
May 30	92	28	1.41″	
May 31	95	28	2.78″	
June 1	104	29	1.24″	
June 2	99	31	0.86″	
June 3	91	28	0.90″	
June 4	92	30	1.00″	
June 5	95	32	1.63″	
June 6	94	30	1.18″	
June 7	91	34	0.94″	
June 8	95	29	1.19″	
June 9	94	32	1.35″	
June 10	93	32	1.12″	
June 11	93	32	2.25″	
June 12	96	37	2.21″	
June 13	94	31	1.40″	
June 14	96	32	2.20″	
June 15	97	35	1.88″	
June 16	98	33	1.80″	
June 17	98	35	3.00″	
June 18	96	37	2.17″	
June 19	98	37	3.44″	
June 20	98	36	1.52″	
June 21	97	30	0.83″	
June 22	93	34	1.78″	
June 23	96	37	1.81″	
June 24	95	33	2.40″	
June 25	97	33	3.50″	
June 26	102	36	1.90″	
June 27	102	38	1.46″	
June 28	101	38	1.19″	
June 29	100	40	1.28″	
June 30	99	40	2.27″	
July 1	103	35	1.70″	
July 2	100	38	0.81″	
July 3	98	42	1.37″	
July 4	97	37	2.49″	
July 5	94	38	4.01″	
July 6	97	31	1.21″	
July 7	**105**	44	2.17″	
July 8	104	43	1.61″	
July 9	102	38	1.33″	
July 10	104	38	1.81″	
July 11	104	41	1.18″	
July 12	95	41	1.00″	
July 13	100	44	2.83″	
July 14	98	43	1.75″	
July 15	96	41	2.10″	
July 16	98	40	1.30″	
July 17	101	43	3.20″	
July 18	98	44	1.07″	
July 19	99	41	1.32″	
July 20	97	41	1.87″	
July 21	96	37	2.24″	
July 22	96	46	2.20″	
July 23	97	42	2.34″	
July 24	101	42	1.07″	
July 25	97	45	3.31″	
July 26	99	45	1.58″	
July 27	99	43	1.60″	
July 28	98	43	1.46″	
July 29	100	43	2.40″	
July 30	102	41	1.50″	
July 31	102	44	1.35″	
August 1	100	40	1.48″	
August 2	98	41	1.40″	
August 3	96	37	1.00″	
August 4	97	40	2.00″	
August 5	100	42	3.40″	
August 6	100	45	1.46″	
August 7	97	41	1.60″	
August 8	98	43	1.55″	
August 9	95	40	2.54″	

Traverse City *(continued)*

	RECORD HIGH	RECORD LOW	RECORD PRECIPITATION	RECORD SNOWFALL
August 10	98	41	1.30″	
August 11	98	41	1.11″	
August 12	95	41	1.53″	
August 13	96	40	3.00″	
August 14	95	41	1.20″	
August 15	97	38	1.53″	
August 16	96	38	1.72″	
August 17	94	40	1.69″	
August 18	100	40	3.00″	
August 19	99	41	4.25″	
August 20	98	38	1.75″	
August 21	99	41	1.27″	
August 22	98	34	2.00″	
August 23	98	39	**4.30″**	
August 24	98	39	1.85″	
August 25	98	37	3.25″	
August 26	98	40	2.32″	
August 27	96	37	1.20″	
August 28	95	36	2.01″	
August 29	96	32	2.02″	
August 30	93	39	1.70″	
August 31	95	39	3.13″	
September 1	96	40	3.93″	
September 2	96	33	1.69″	
September 3	94	37	1.47″	
September 4	92	40	1.37″	
September 5	95	38	1.65″	
September 6	92	37	1.48″	
September 7	93	37	2.15″	
September 8	95	37	1.43″	
September 9	94	34	0.86″	
September 10	96	33	1.57″	
September 11	95	32	2.55″	
September 12	95	31	1.23″	
September 13	93	28	2.36″	
September 14	94	30	1.81″	
September 15	96	30	1.44″	
September 16	90	32	1.45″	
September 17	90	32	1.30″	
September 18	90	32	2.49″	
September 19	88	35	1.33″	
September 20	90	32	1.78″	
September 21	88	33	2.20″	
September 22	90	31	1.56″	
September 23	90	29	2.62″	
September 24	89	30	3.06″	
September 25	89	27	1.34″	
September 26	88	31	2.10″	
September 27	88	27	0.75″	
September 28	85	27	2.10″	
September 29	90	28	2.31″	0.5″

	RECORD HIGH	RECORD LOW	RECORD PRECIPITATION	RECORD SNOWFALL
September 30	89	26	3.75″	
October 1	88	30	1.75″	
October 2	89	30	1.07″	1.9″
October 3	87	27	2.01″	
October 4	88	28	1.40″	
October 5	87	25	1.00″	
October 6	87	24	1.80″	0.5″
October 7	88	25	1.61″	
October 8	88	29	0.80″	
October 9	80	25	0.60″	0.5″
October 10	86	25	1.42″	1.0″
October 11	85	23	1.15″	2.0″
October 12	84	26	1.26″	
October 13	83	26	2.11″	
October 14	85	25	0.78″	
October 15	85	26	1.25″	
October 16	85	23	1.12″	0.7″
October 17	85	22	1.50″	2.8″
October 18	83	15	1.50″	1.5″
October 19	80	24	1.10″	1.2″
October 20	79	22	1.29″	0.2″
October 21	82	21	1.10″	
October 22	83	22	1.02″	
October 23	82	22	1.36″	1.0″
October 24	80	24	1.92″	1.0″
October 25	77	23	1.58″	1.4″
October 26	76	20	0.74″	1.9″
October 27	81	23	1.48″	1.0″
October 28	80	20	0.76″	2.0″
October 29	75	22	1.00″	0.2″
October 30	78	22	0.61″	2.0″
October 31	76	22	0.60″	0.1″
November 1	75	23	2.56″	6.0″
November 2	77	15	1.01″	1.5″
November 3	78	9	0.63″	10.1″
November 4	76	13	1.34″	4.0″
November 5	71	14	1.32″	6.0″
November 6	74	2	1.39″	4.5″
November 7	70	14	1.33″	8.5″
November 8	72	15	0.96″	2.1″
November 9	71	18	1.42″	2.5″
November 10	66	15	1.44″	3.0″
November 11	69	15	1.16″	9.6″
November 12	69	10	0.90″	9.0″
November 13	64	11	1.22″	5.0″
November 14	65	7	1.13″	4.0″
November 15	67	7	2.50″	5.0″
November 16	71	11	0.90″	4.0″
November 17	70	6	0.85″	4.0″
November 18	69	11	1.55″	3.0″

Traverse City *(continued)*

	RECORD HIGH	RECORD LOW	RECORD PRECIPITATION	RECORD SNOWFALL
November 19	69	14	1.20″	12.0″
November 20	65	2	0.92″	7.0″
November 21	68	11	1.30″	3.9″
November 22	64	11	1.28″	6.0″
November 23	66	7	1.20″	3.0″
November 24	64	–5	0.45″	3.0″
November 25	60	–5	1.45″	10.0″
November 26	63	3	1.45″	6.0″
November 27	66	3	2.46″	10.0″
November 28	62	6	1.50″	7.2″
November 29	66	5	1.60″	**16.0**″
November 30	65	4	1.10″	5.0″
December 1	63	1	1.30″	6.6″
December 2	62	3	0.62″	7.8″
December 3	64	0	1.22″	9.0″
December 4	61	–5	0.43″	6.0″
December 5	63	2	1.40″	6.0″
December 6	56	–1	0.80″	8.0″
December 7	57	–4	1.29″	8.0″
December 8	60	–3	0.95″	5.0″
December 9	60	1	0.55″	6.0″
December 10	54	–6	1.46″	6.1″
December 11	58	–11	1.30″	6.0″
December 12	58	–8	0.70″	6.0″
December 13	55	–4	0.70″	7.0″
December 14	56	–5	0.72″	5.0″
December 15	54	–6	1.21″	8.1″
December 16	60	–3	1.18″	14.4″
December 17	56	–10	0.50″	6.1″
December 18	49	–3	0.62″	6.0″
December 19	59	–9	0.36″	5.0″
December 20	55	–13	0.75″	5.0″
December 21	55	–4	1.68″	4.0″
December 22	55	–2	0.60″	6.0″
December 23	51	–6	1.61″	7.0″
December 24	48	–7	0.60″	6.0″
December 25	58	–6	0.87″	7.4″
December 26	48	–5	0.90″	6.6″
December 27	53	–11	1.25″	8.0″
December 28	59	–8	0.95″	10.0″
December 29	59	–26	0.46″	6.0″
December 30	56	–13	1.00″	8.8″
December 31	57	–5	1.00″	9.0″

Bibliography

Associated Press. "Global Warming May Pose Threat to Heart." Vienna, Austria, September 5, 2007.

Astronomy Magazine. www.astronomy.com.

Badgley, Catherine, John C. Barry, Michèle E. Morgan, Sherry V. Nelson, Anna K. Behrensmeyer, Thure E. Cerling, and David Pilbeam. "Ecological Changes in Miocene Mammalian Record Show Impact of Prolonged Climatic Forcing." *Proceedings of the National Academy of Sciences of the United States of America* 105, no. 34 (2008): 12145–49.

"Better Holiday Lights: Get That Warm Glow with Less Energy." *Consumer Reports*, December 2007.

Bohnak, Karl. *So Cold a Sky: Upper Michigan Weather Stories*. Negaunee, Mich.: Cold Sky Publishing, 2006.

Bortkiewicz, Alicja, Elżbieta Gadzicka, Wiesław Szymczak, Agata Szyjkowska, Wiesława Koszada Włodarczyk, and Teresa Makowiec-Dąbrowska. "Physiological Reaction to Work in Cold Microclimate." *International Journal of Occupational Medicine and Environmental Health* 19, no. 2 (2006): 123–31.

Clark, P. U., A. J. Weaver, E. Brook, E. R. Cook, T. L. Delworth, and K. Steffen. *CCSP, 2008: Abrupt Climate Change. A Report by the U.S. Climate Change Science Program and the Subcommittee on Global Change Research*. U.S. Geological Survey, December 16, 2008.

Climate Change Science for Broadcast Meteorologists and Weathercasters. Conference held in Chicago, Ill., April 2009.

Collections and Researches Made by the Michigan Pioneer and Historical Society. Wynkoop Hallenbeck Crawford Company, State Printers, 1901.

Collections Report of the Pioneer Society of the State of Michigan Together with Reports of County Pioneer Societies. Wynkoop Hallenbeck Crawford Company, State Printers, 1908.

Cooperative Program for Operational Meteorology, Education and Training

(COMET). *Climate Change: Fitting the Pieces Together.* Boulder, Colo.: University Corporation for Atmospheric Research, 2009.

Diffenbaugh, Noah S., Christian H. Krupke, Michael A. White, and Corrine E. Alexander. *Global Warming Presents New Challenges for Maize Pest Management.* Purdue University, December 16, 2008.

Doesken, Nolan. "Let It Rain." *Weatherwise* 60 (July–August 2007): 50–55.

Early American Tornadoes: 1586–1870. American Meteorological Society, 1970.

Earth Gauge: Lightning Know-How. National Environmental Education Foundation Program, June 2, 2008.

Earth System Research Laboratory. http://www.esrl.noaa.gov.

Frisinger, H. Howard. *The History of Meteorology to 1800.* American Meteorological Society, 1983.

Future Retreat of Arctic Sea Ice Will Lower Polar Bear Populations and Limit Their Distribution. United States Department of the Interior, United States Geologic Survey, September 7, 2007.

Grazulis, Thomas P. *Significant Tornadoes.* St. Johnsbury, Vt.: Environmental Films, 1990.

Great Lakes Information Network. http://www.great-lakes.net/lakes/#overview.

Hamilton, Calvin J. Views of the Solar System (Web site). http://www.solarviews.com.

Hoenisch, B., N. G. Hemming, J. F. McManus, D. Archer, and M. Siddall. "Atmospheric Carbon Dioxide Concentration across the Mid-Pleistocene Transition." *Science,* June 19, 2009, 1551–54.

Huff, Floyd A., and James R. Angel. *Rainfall Frequency Atlas of the Midwest.* Bulletin 71. Illinois State Water Survey, Champaign, Ill., 1992.

Humane Society of the United States. http://www.hsus.org.

Intergovernmental Panel on Climate Change. *Climate Change 2007: Synthesis Report. Contribution of Working Groups I, II and III to the Fourth Assessment Report of the Intergovernmental Panel on Climate Change.* Geneva, Switzerland.

Intergovernmental Panel on Climate Change. www.ipcc.ch.

Knutson, Thomas R., John L. McBride, Johnny Chan, Kerry Emanuel, Greg Holland, Chris Landsea, Isaac Held, James P. Kossin, A. K. Srivastava, and Masato Sugi. "Tropical Cyclones and Climate Change." *Nature Geoscience,* no. 3 (2010): 157–63.

Lam, Tina, and Chris Christoff. "Deadly Chain Reaction." *Detroit Free Press,* August 27, 2007.

Lichter, S. Robert. *Climate Scientists Agree on Warming, Disagree on Dangers, and Don't Trust the Media's Coverage of Climate Change.* Statistical Assessment Service, George Mason University, April 24, 2008.

Ludlum, David. *Early American Winters: 1604–1820.* American Meteorological Society, 1966.

Lüthi, D., M. Le Floch, B. Bereiter, T. Blunier, J. Barnola, U. Siegenthaler, D. Raynaud, J. Jouzel, H. Fischer, K. Kawamura, and T. Stocker. "High-Resolution Carbon Dioxide Concentration Record 650,000–800,000 Years before Present." *Nature* 453 (March 2008): 379–82.

Meehl, Gerald A., Claudia Tebaldi, Guy Walton, David Easterling, and Larry McDaniel. "The Relative Increase of Record High Maximum Temperatures Compared to Record Low Minimum Temperatures in the U.S." *Geophysical Research Letters* 36, L23701, doi:10.1029/2009GL040736, December 1, 2009.

Michigan Committee for Severe Weather Awareness. http://mcswa.org.

Midwest Energy Efficiency Alliance. http://www.mwalliance.org.

Myers, Philip, Barbara L. Lundrigan, Susan M. G. Hoffman, Allison Poor Haraminac, and Stephanie H. Seto. "Climate-Induced Changes in the Small Mammal Communities of the Northern Great Lakes Region." *Global Change Biology* 15 (June 2009): 1434–54.

National Climatic Data Center. www.ncdc.noaa.gov.

National Oceanic and Atmospheric Administration, Office of Climate, Water and Weather Services. http://www.weather.gov/om.

National Oceanic and Atmospheric Administration, Public Affairs Office. "History of the National Weather Service." http://www.weather.gov/pa/history/index.php.

National Snow and Ice Data Center. Press release, October 1, 2007. http://nsidc.org/news/press/2007_seaiceminimum/20071001_pressrelease.html.

National Snow and Ice Data Center. Press release, October 6, 2009. http://nsidc.org/news/press/20091005_minimumpr.html.

National Weather Service, Buffalo, N.Y. www.erh.noaa.gov/buf.

National Weather Service, Detroit/Pontiac, Mich. www.crh.noaa.gov/dtx.

National Weather Service, Gaylord, Mich. www.crh.noaa.gov/apx.

National Weather Service, Grand Rapids, Mich. www.crh.noaa.gov/grr.

National Weather Service, Marquette, Mich. www.crh.noaa.gov/mqt.

National Weather Service, National Hurricane Center. http://www.nhc.noaa.gov.

National Weather Service, Northern Ind. www.crh.noaa.gov/iwx.

National Weather Service, St. Louis, Mo. www.crh.noaa.gov/lsx.

National Weather Service Storm Prediction Center. http://www.spc.noaa.gov/products.

Rifkin, Jeremy. "The Crises under the Ice." *Los Angeles Times*, August 9, 2007.

"Rising Acidity Levels Could Trigger Shellfish Revenue Declines, Job Losses." Woods Hole Oceanographic Institution, Media Relations Office, June 17, 2009.

Rundquist, Larry A. *Ice Breakup Reconnaissance Approach in Alaska*. National Weather Service River Forecast Center.

Sillars, Malcolm P. *1989 Detroit Weather Book.* Mal Sillars Weather Consultants, 1988.

Solomon, Susan, Gian-Kasper Plattner, Tero Knutti, and Pierre Friedlingstein. "Irreversible Climate Change Due to Carbon Dioxide Emissions." *Proceedings of the National Academy of Sciences of the United States of America* 106, no. 6 (2009): 1704–9.

Sousounis, Peter, and Patty Glick. *The Potential Impacts of Global Warming on the Great Lakes Region.* First National Assessment of the Potential Consequences of Climate Variability and Change. United States Environmental Protection Agency, Global Change Research Program, 2000.

Union of Concerned Scientists and Ecological Society of America. *Confronting Climate Change in the Great Lakes Region.* 2005.

United States Department of Commerce/National Oceanic and Atmospheric Administration/National Weather Service. *Thunderstorms . . . Tornadoes . . . Lightning . . . Nature's Most Violent Storms.* http://www.nws.noaa.gov/om/brochures/ttl.pdf.

United States Environmental Protection Agency. www.epa.gov.

U.S. Army Corps of Engineers. *Historic Great Lakes Levels.* http://www.lre.usace.army.mil/greatlakes/hh/greatlakeswaterlevels/historicdata/greatlakeshydrographs.

van Mantgem, Phillip J., Nathan L. Stephenson, John C. Byrne, Lori D. Daniels, Jerry F. Franklin, Peter Z. Fulé, Mark E. Harmon, Andrew J. Larson, Jeremy M. Smith, Alan H. Taylor, and Thomas T. Veblen. "Widespread Increase of Tree Mortality Rates in the Western United States." *Science,* January 22, 2009, 521–24.

Index

Adams, Linda, x
air masses, 5, 43
air pressure, 6
Alberta clippers, 28
Almagest, 2
American Geophysical Union, 59
American Meteorological Society, 58
apparent temperature. *See* heat index; wind chill
Arctic Oscillation, 7
Aristotle, 1–2
Armistice Day Storm, 37
aurora australis, 70
aurora borealis, 70
average temperatures, 11–12

barometer, 6
Beachler, Dave, x
Berger, Dave, x
black ice, 67
blizzard warning, 33

cabin fever, 33
Campbell, Gary, x
carbon dioxide. *See* global warming
CFL bulbs, 61–62
Charlevoix Public Library, x
chroma-key, 71
climate change. *See* global warming
clouds, lake effect, 33
CoCoRaHS. *See* Community Collaborative Rain, Hail and Snow network
cold. *See* temperature
cold, dressing for, 46
coldest temperature ever in Michigan, 44
cold fronts, 5
comets, 68–69
Community Collaborative Rain, Hail and Snow network, 74
computer models, 8–10, 31
  climate models, 53–54
conservation of angular momentum, 17
convergence, 5
cooperative weather observers, 72, 74
Crupi, Kevin, x

Deedler, Bill, x, 64
derecho, 37
dewpoint temperature, 4, 44
Doppler radar, 18, 22
dynamics, 4–5

Earth's tilt, 2–3
Edmund Fitzgerald, 29
El Niño, 7, 51, 55
Erwin, Mary, x
evaporation, 9,16, 31, 35, 41–42, 45, 56

floods, 35–37
  ice jams, 36
  of 1986, 35–36
  safety, 38
  seiches, 37–38
  warnings, 38
fog, 66–67
forecasting, 7–10, 26–27, 31–32, 36, 44
Fox, Adam, x
freezing rain, 33–35

fronts, 5–6
frostbite, 45–46
Fujita Tornado Intensity Scale, 19

Gaidica, Chuck, x–xi
Gales of November, 27–29, 37
  Armistice Day Storm, 37
  1913 "White Hurricane," 27–29
global warming, 12, 48–63
  Arctic ice, 53–54, 58
  carbon dioxide, 49–52, 54, 63
  "climategate," 59–60
  deuterium (in ice cores), 52
  effects on Earth, 53–54
  effects on Michigan, 55–57
  El Niño, 51, 55
  greenhouse effect, 48–49
  greenhouse gases, 49–53
  ice cores, 51–52
  La Niña, 55
  methane, 49, 51, 54
  nitrous oxide, 49
  paleoclimatic time line, 51–52
  permafrost, 54–55
  polar ice, 51, 53–54
  tree rings, 51
  uncertainties, 55
  urban heat island effect, 52
Grant, President Ulysses S., 70
Grazulis, Tom, 19
Great Lakes, 34–35
  temperatures, 18, 27, 31
Great Lakes Aggregate, 27
Great Lakes Storm of 1913, 27–29
Greely, Major General Adolphus, 71
greenhouse effect. *See* global warming
greenhouse gases. *See* global warming
green screen. *See* chroma-key
Gross, Adam, xi
Gross, Jared, xi, 23
Gross, Marion, xi
Gross, Marvin, xi
Gross, Nancy, xi
gustnadoes, 18

hail, 15–16, 75
  Aristotle theory, 1–2
Harrington, Mark W., 71
heat. *See* temperature
heat advisory, 42
heat exhaustion, 42
heat index, 41–42
heat stroke, 42–43
Heat Wave of 1936, 40–41
high pressure, 6
Holsten, Todd, x
hottest temperature ever in Michigan, 41
humidity, 4, 41–42, 67
hurricanes, 58, 64–66
hydrostatic equation, 40
hypothermia, 46–47

ice. *See* black ice; freezing rain
ice jams, 36
ice pellets, 32
ice storms, 32–33
ideal gas law, 3, 40
Intergovernmental Panel on Climate Change (IPCC), 50–53, 59–60
Ivan, Judith, x

jet stream, 6–7, 27

Keysor, James, x
Kompoltowicz, Keith, x
Korzeniewski, Bryant, x

La Niña, 7, 55
lake breeze, 68
lake-effect clouds, 33
lake-effect snow, 30–32
Lashley, Sam, x
lightning, 14–15
Lothamer, Brentley, x
low pressure, 6

Madden/Julian Oscillation, 7
Marino, Bill, x
Mars, 49
McCallum, Sam, x
McCarthy, Ellen, x
MCSWA. *See* Michigan Committee for Severe Weather Awareness
Meehl, Jerry, x, 50
Mercury, 49
Meteorologica, 1–2

meteorological instruments, development of, 2
meteors, 69–70
methane. *See* global warming
Miami (Ohio) University, 55
Michigan Committee for Severe Weather Awareness, 25
Michigan State University, 55–56
Milton, Chris, x
Montgomery, Brian, x, 9
Moon, 6, 50

National Climatic Data Center, ix, x, 72, 74
National Weather Service, ix, 9, 16, 24, 33, 36, 38, 42, 45–46, 64, 70–71, 72, 74
NCDC. *See* National Climatic Data Center
nitrous oxide. *See* global warming
normal weather, 11–12
North Atlantic Oscillation, 7
northern lights. *See* aurora borealis
Nugent, James, 56
NWS. *See* National Weather Service

Passel, Charles, 45
Pennoyer, H., 39
Pollman, Rich, x
precipitable water, 31
pressure. *See* air pressure
Ptolemy, 2
Purdue University, 56

radar, 32
  Doppler radar, 18, 22
radiation. *See* solar radiation and output
radiational cooling, 43–44
radiosondes (weather balloons), 8, 32
rain (or raindrops), 4, 16, 18, 27, 31–32, 35–36, 57, 65–67
rainbows, 66
rising air, 4–6

Samenow, Jason, x
satellites, 32
Schenck-Gustafsson, Karin, 56
seasonal affective disorder, 33
seiches, 37–38
severe thunderstorms, wind and hail criteria, 15–17
Shein, Karsten, x
Shey, Sam, x
shooting stars. *See* meteors
Significant Tornadoes, 19
Sillars, Mal, x–xi, 9, 27
Siple, Major Paul, 45
sleet, 32
snow
  early season, 29
  effect on lake levels, 34–35
  forecasting, 30–32
  lake-effect, 30, 32
  snow-to-water ratio, 31
  synoptic, 31
  thundersnow, 31
solar radiation and output, 3, 49, 54, 67
solar wind, 69
Somerville, Richard, 50
special marine warning, 24
supercell, 17–18

temperature
  aloft, 7–8
  Aristotle theory, 1–2
  average, 11–12
  "Bummer Summer of 1992," 67–68
  coldest ever in Michigan, 44
  dewpoint temperature, 4, 44
  European heat wave of 2003, 56
  extreme cold, 43–45
  extreme heat, 40–41
  global warming impact. *See* global warming
  Heat Wave of 1936, 40–41
  highest ever in Michigan, 41
  in ideal gas law, 7
  impact on freezing rain, 32–33
  impact on rising air, 10
  impact on snow, 31
  lake breeze impact, 68
  normal, 11–12
  planet's average, 49, 55
  in vehicles, 43
thunder, 14–15
thundersnow, 31
thunderstorms, vii, 5, 8, 14–19, 22, 35, 37, 68

tornadoes
  detection by Doppler radar, 18, 22
  development of, 17–18
  gustnadoes, 18
  in Michigan, 18–22
  safety, 22–25
  sirens, 23
  Tornado Alley, 18–19
  tornado warnings, 22–23
  tornado watches, 22
  waterspouts, 18
training (continuous waves of thunderstorms), 35
Trenberth, Kevin, 50
triggers, 5

University of East Anglia Climate Research Unit, 59
University of Michigan, 9
urban heat island, 52
urban sprawl, 21, 36
upper air data, 8
  mandatory levels, 8
  radiosondes (weather balloons), 8, 32
U.S. Army Corps of Engineers, x
U.S. Geological Survey (USGS), 57–58

Venus, 49
volcanoes (impact of eruptions on Michigan), 67–68

Walton, Mark, x
Warfield, Bob, x–xi
warm fronts, 6
waterspouts, 18
water vapor, 4, 31, 41, 49, 66
weather balloons. *See* radiosondes
Weather Bureau, 71
weather forecasting, 7–10
  accuracy, 9
  computer models, 8
  wrong forecasts, 9–10
wind
  aloft, 7
  Aristotle theory, 1
  associated with fronts, 5
  caused by pressure change, 6
  Doppler radar detection of, 18
  in *Edmund Fitzgerald* storm, 29
  hurricane remnants hitting Michigan, 64–65
  impact on cold temperatures, 43–44
  impact on lake-effect snow, 30
  impact on seiches, 37–38
  impact on tornado formation, 17–18
  impact on wind chill, 45
  lake breeze, 68
  in 1913 "White Hurricane" storm, 28–29
  relationship to weather forecasting, 7
  severe, 16–17
  straight-line, 17
wind chill, 45–46
  advisory, 46
  warning, 46
winter storm warning, 33
winter storm watch, 33
winter weather advisory, 33

Zika, Matt, x